高职高专机电类专业系列教材

机电专业英语

主　编　谢　敏　周爱华
副主编　郭　燕
参　编　王寿斌　秦培亮
主　审　楼佩煌

机械工业出版社

本书由机械制造基础、机床、机电一体化基础、机电一体化技术、工程应用和文献检索6个部分组成，分成26个单元，包括：工程材料的分类，极限尺寸、配合与公差，金属的热处理，成形，车床，铣床，磨削分类与无心磨削，铣削，钻头与钻床，电工技术，电子技术，自动控制系统，机电一体化系统，AT89S51介绍，西门子可编程序控制器介绍，液压技术，气动技术，电气工程，数控技术简介，WinCC flexible介绍，CAD/CAM/CAPP，IRB 140工业机器人数据手册，机器视觉，自动装配线，信息检索简介，专业文献的检索等。

本书可以作为高职高专院校机电类、自动化类等专业的专业英语教材，也可作为相关专业技术人员和行业英语爱好者的学习参考书。

图书在版编目（CIP）数据

机电专业英语/谢敏，周爱华主编. —北京：机械工业出版社，2012.8
（2024.7重印）
高职高专机电类专业系列教材
ISBN 978-7-111-38758-9

Ⅰ. ①机… Ⅱ. ①谢…②周… Ⅲ. ①机电工业—英语—高等职业教育—教材 Ⅳ. ①H31

中国版本图书馆CIP数据核字（2012）第169208号

机械工业出版社（北京市百万庄大街22号　邮政编码100037）
策划编辑：于　宁　责任编辑：于　宁　王宗锋　版式设计：霍永明
责任校对：樊钟英　封面设计：陈　沛　　　　　责任印制：常天培
北京机工印刷厂有限公司印刷
2024年7月第1版第4次印刷
184mm×260mm · 12.75印张 · 309千字
标准书号：ISBN 978-7-111-38758-9
定价：39.80元

电话服务　　　　　　　　　　　网络服务
客服电话：010-88361066　　　机　工　官　网：www.cmpbook.com
　　　　　010-88379833　　　机　工　官　博：weibo.com/cmp1952
　　　　　010-68326294　　　金　书　网：www.golden-book.com
封底无防伪标均为盗版　　　机工教育服务网：www.cmpedu.com

前　言
PREFACE

　　为了适应高等职业教育的不断发展，针对高职高专院校机电类、自动化类专业学生的培养目标和岗位技能要求，在充分体现理论内容"必须、够用"的原则和突出应用能力及综合素质培养的前提下，本书通过专业英语和专业课程内容的相互融合，介绍了本专业的基本理论、应用技术、产品及应用软件的英文陈述与表达和机电一体化技术的前沿应用。

　　本书以现代机电一体化技术为背景，在教材中充分体现专业特色，尽可能兼顾机电专业的各个技术侧面，对机电一体化技术专业涉及的理论知识与技术、产品、应用软件进行了有侧重的选择和精简，并循序渐进、由浅入深、由简到繁地进行了陈述，使读者能得到较为全面的现代机电技术专业英语的阅读、词汇和知识扩充。全书由6部分组成，包括机械制造基础、机床、机电一体化基础、机电一体化技术、工程应用和文献检索，共分成26个单元，每个单元由课文、单词与词组、课文注释、练习等组成。书后附有课文译文及练习答案、常用专业词汇。

　　本书力求做到内容正确、全面，从理论基础到技术应用和产品介绍，各单元安排和内容组成充分考虑了专业基础技术、工程实际应用技术和最新技术的实际特点。

　　本书由南京化工职业技术学院谢敏、甘肃机电职业技术学院周爱华、南京化工职业技术学院郭燕、苏州工业园区职业技术学院王寿斌和苏州农业职业技术学院秦培亮编写。本书的编写分工是：谢敏编写 Unit 13、14、19、20、22、23、24 以及 Part Ⅳ 的阅读材料 A、B 和 Part Ⅴ 的阅读材料 A～D；周爱华编写 Unit 1～4 及 Part Ⅰ 的阅读材料 A～D，Unit 5～8 及 Part Ⅱ 的阅读材料 A～D；郭燕编写 Unit 9～12 及 Part Ⅲ 中的阅读材料 A～D；王寿斌编写 Unit 21、25、26；秦培亮编写 Unit 15～18 及 Part Ⅳ 的阅读材料 C、D。谢敏、周爱华担任主编，负责全书的统稿工作。南京航空航天大学楼佩煌教授担任主审，他对本教材提出了许多宝贵的意见和建议。

　　本书在编写时，参考了大量的文献资料及国外公司的技术资料，在此向这些文献的作者深表谢意。

　　由于编者水平有限，书中难免存在不足之处，恳请读者批评指正。

编　者

目录 CONTENTS

前言

Part Ⅰ Fundamentals of Manufacturing 1
  Unit 1   Classification of Materials 1
  Unit 2   Limits, Fits and Tolerances 4
  Unit 3   Heat Treatment of Metal 7
  Unit 4   Forming 9
  Reading Material A   Kinds of Steel 12
  Reading Material B   Dimension and Tolerance 13
  Reading Material C   Metal Hot Working 15
  Reading Material D   Surface Roughness 16

Part Ⅱ Machine Tools 18
  Unit 5   Lathes 18
  Unit 6   Grinding Classification and Centreless Grinding 21
  Unit 7   Milling 24
  Unit 8   Drills and Drill Presses 27
  Reading Material A   Spur and Helical Gears 29
  Reading Material B   Basic Machining Techniques 30
  Reading Material C   Machine Elements 31
  Reading Material D   Milling Machines 32

Part Ⅲ Basics of Mechatronics 35
  Unit 9   Electrician Technology 35
  Unit 10   Electron Technology 37
  Unit 11   Automatic Control Systems 40
  Unit 12   Mechatronics Systems 43
  Reading Material A   Alternating Current 46
  Reading Material B   Digital Television 47
  Reading Material C   Adaptive Control Systems 47
  Reading Material D   Prospect of New Mechatronics Products 48

Part Ⅳ Mechatronics Technique 50
  Unit 13   Introduction to AT89S51 50
  Unit 14   Introduction to SIEMENS PLC 55
  Unit 15   Technology of Hydraulic Pressure 59
  Unit 16   Pneumatic Technology 64

Unit 17　Electrical Engineering ··· 68
Unit 18　Introduction to NC ··· 70
Reading Material A　Power Saving Modes of Microcontrollers ···························· 75
Reading Material B　Guidelines for Designing a Micro PLC System ······················· 76
Reading Material C　Hydraulic Power Transmission ·· 78
Reading Material D　Tooling for Computer Numerical Control Machines ·················· 78

Part V　Applications of Engineering ·· 81
Unit 19　Introduction to WinCC Flexible ··· 81
Unit 20　Flexible Manufacturing Systems ·· 84
Unit 21　CAD/CAM/CAPP ··· 88
Unit 22　Datasheet of IRB 140 Industrial Robot ··· 91
Unit 23　Machine Vision ··· 95
Unit 24　Automatic Assembly ··· 99
Reading Material A　Selecting the Communications Protocol for Your Network ········· 103
Reading Material B　RobotStudio Overview ··· 105
Reading Material C　Computer-Integrated Manufacturing System（CIMS） ············ 106
Reading Material D　Virtual Reality ·· 107

Part VI　Document Retrieval ··· 109
Unit 25　Introduction to Information Retrieval ·· 109
Unit 26　Professional Special Field Document Retrieval ······································ 112

Appendix ·· 116
Appendix A　课文译文及练习答案 ·· 116
Appendix B　常用专业词汇 ··· 180

参考文献 ··· 195

Part I Fundamentals of Manufacturing

Unit 1 Classification of Materials

Materials may be grouped in several ways. Scientists often classify materials by there state: solid, liquid, or gas. They also separate them into organic (once living) and inorganic (never living) materials. For industrial purposes, materials are divided into engineering materials or nonengineering materials. Engineering materials are those used in manufacture and become parts of products. Nonengineering materials are the chemicals, fuels, lubricants, and other materials used in the manufacturing process which do not become part of the product.

Engineering materials may be classified into four groups: metals, ceramics, polymers, and composite materials.

1. Metals

Metals are generally defined as those elements whose hydroxides form bases (such as sodium or potassium). Metals may exist as pure elements. When two or more metallic elements are combined, they form a mixture called an alloy.

The term alloy is used to identify any metallic system. In metallurgy, it is a substance with metallic properties, that is composed of two or more elements, intimately mixed. Of these elements, one must be a metal. Plain carbon steel, in the sense, is basically an alloy of iron and carbon. Other elements are present in the form of impurities. However, for commercial purposes, plain carbon steel is not classified as an alloy steel.

Metals and alloys, which include steel, aluminum, magnesium, zinc, cast iron, titanium, copper, nickel, and many others, have the general characteristics of good electrical and thermal conductivity, relatively high strength, high stiffness, ductility or formability, and shock resistance. They are particularly useful for structural or load-bearing applications. Although pure metals are occasionally used, alloys are normally designed to provide improvement in a particular desirable property or permit better combinations of properties.

2. Ceramics

Ceramics, such as brick, glass, tableware, insulators, and abrasives, have poor electrical and thermal conductivity. Although ceramics may have good strength and hardness, their ductility, formability, and shock resistance are poor. Consequently, ceramics are less often used for structural or load-bearing applications than metals. However, many ceramics have excellent resistance to high temperatures and certain corrosive media and have a number of unusual and desirable optical, electrical, and thermal properties.

3. Polymers

Polymers include rubber, plastics, and many types of adhesives. They are produced by creating large molecular structures from organic molecules, obtained from petroleum or agricultural products, in a process known as polymerization. Polymers have poor electrical and thermal conductivity, low strengths, and are not suitable for use at high temperatures. Some polymers have excellent ductility, formability, and shock resistance while others have the opposite properties. Polymers are lightweight and frequently have excellent resistance to corrosion.

4. Composite Materials

Composite materials are formed from two or more materials, whose properties cannot be obtained by any single material. Concrete, plywood, and fiberglass are typical, although crude, examples of composite materials. With composite materials, we can produce lightweight, strong, ductile, high heat-resistant materials that are otherwise unobtainable, or produce hard yet shock resistant cutting tools that would otherwise shatter.

New Words and Phrases

1. lubricant ['luːbrikənt] n. 润滑剂
2. ceramic [si'ræmik] n. 陶瓷，陶瓷制品
3. polymer ['pɔlimə] n. 聚合物
4. composite ['kɔmpəzit] adj. 合成的，复合的，混合物
5. sodium ['səudiːəm] n. 钠
6. potassium [pə'tæsiːəm] n. 钾
7. zinc [ziŋk] n. 锌
8. cast iron 铸铁
9. titanium [tai'teiniəm] n. 钛
10. nickel ['nikəl] n. 镍
11. stiffness ['stifnis] n. 硬度，刚度
12. ductility [dʌk'tiliti] n. 韧性，可延展性
13. formability ['fɔːmə'biliti] n. 可成型性
14. shock resistance 抗冲击性
15. load-bearing 承载
16. tableware ['teib(ə)lweə] n. 餐具
17. abrasive [ə'breisiv] n. 研磨剂，磨料（具）；adj. 研磨的
18. thermal conductivity 导热性
19. adhesive [æd'hiːsiv] n. 黏结剂，黏胶剂
20. molecular [məu'lekjuːlə] adj. 分子的
21. molecule ['mɔlikjuːl] n. 分子
22. polymerization [,pɔlimərai'ziʃən] n. 聚合
23. corrosion [kə'rəuʒən] n. 腐蚀
24. plywood ['plaiwud] n. 夹板

25. fiberglass ['faibəglɑːs] n. 玻璃纤维
26. crude [kruːd] adj. 天然的，未经加工的
27. heat-resistant material　耐热材料
28. shatter ['ʃætə] v. 粉碎，破坏
29. be used for...　用于……
30. be formed from...　由……组成

Notes

1. Consequently, ceramics are less often used for structural or load-bearing applications than metals.

因此，与金属相比，陶瓷很少用于结构件或承载件。

less... than = not so... as　比……少

be used for...　是"用于……"的

例如：A hammer is used for driving in nails。

2. They are produced by creating large molecular structures from organic molecules, obtained from petroleum or agricultural products, in a process known as polymerization.

它们是由来自石油或农产品的有机分子通过聚合形成的巨大分子结构所产生的。

3. Composite materials are formed from two or more materials, whose properties cannot be obtained by any single material.

复合材料由两种或两种以上的材料组成，其性能绝非任何一种单一材料所能拥有。

be formed from...　由……组成

"whose properties cannot be obtained by any single material" 的 "whose" 指的是复合材料。

Exercises

1. Answer the following questions according to the text.

a) What is metal? Try to describe it using the words in this passage or of your own.

b) What is polymer? Try to describe it using the words in this passage or of your own.

2. Decide whether the following statements are true (T) or false (F) and put "T" or "F" in the brackets according to the text.

a) Materials are classified into five groups: metals, nonmetals, ceramics, polymers, and composite materials. (　　)

b) Metals and alloys have relatively high strength, high stiffness, ductility or formability, but low shock resistance. (　　)

c) Ceramics have poor electrical and thermal conductivity, may have high strength and hardness, and have high ductility, formability, and shock resistance. (　　)

d) Polymers are lightweight and frequently have poor resistance to corrosion. (　　)

Unit 2　Limits, Fits and Tolerances

Terminology

Fit: The relation resulting from the difference between the sizes of two mating parts.

Basic Size: The size with reference to which the limits of size are fixed; also termed the nominal size or nominal dimension.

Actual Size: The size of a part as may be found by measurement.

Limits of Size: The two extreme permissible sizes between which the actual size is contained. The two extreme sizes are termed the maximum limit and minimum limit.

Tolerance: The difference between the maximum limit and the minimum limit. (The tolerance is also equal to the algebraic difference between the upper and lower deviations.)

Upper Deviation: The algebraic difference between the maximum limit of basic size and the corresponding basic size.

Lower Deviation: The algebraic difference between the minimum limit of basic size and the corresponding basic size.

Types of Fit

Depending upon the actual limits of the hole or shaft, a fit may be classified as follows.

(1) **Clearance Fit**: A fit that always provides a clearance between the mating parts. In this case, the tolerance zone of the hole is entirely above that of the shaft.

(2) **Interference Fit**: A fit that always provides an interference between the mating parts. Here, the tolerance zone of the hole is entirely below that of the shaft.

(3) **Transition Fit**: A fit that may provide, depending on the actual dimensions of the finished products, either a clearance or an interference between the mating parts.

Of the various methods used to apply the system of fits, the principal ones are the shaft-basis system and hole-basis system. In the shaft-basis system, the different clearances and interferences are obtained by associating various holes with a single basic size of shaft, the upper deviation of which is zero (symbol h). In the hole-basis system, the different clearances and interferences are obtained by associating various shafts with a single basic size of hole, the lower deviation of which is zero (symbol H).

Normally, it is easier to produce a shaft with a specified tolerance than a hole with the same tolerance. Consequently, in modern engineering design, the hole-basis system is most extensively used and our discussion will refer mostly to this system. However, the designer should decide on the adoption of either system to enable general interchangeability.

Symbols for Tolerances and Fits

A tolerance is designated by a letter (in some cases, two letters), a symbol, and a numerical symbol. Capital letters are used for holes and small letters for shafts. The letter symbol indicates the position of the zone of tolerance in relation to the zero line representing the basic size. The numerical symbol represents the value of this zone of tolerance and is called the grade or quality of tolerance.

Both the position and the grade of tolerance are functions of the basic size. A toleranced size is thus defined by its basic value followed by a letter and a number, e. g.

$$\phi 50H7 \quad \text{or} \quad \phi 50g6$$

Fundamental Aspects of Tolerance System

For all industrial measurements, the standard reference temperature is 293K. Basic sizes from 1mm to 500mm have been subdivided into 13 steps or ranges. From 500mm to 3150mm, there are 8 nominal steps. For each nominal step, there are 20 grades of tolerance designated IT01, IT0, and IT1,...... IT17, IT18. These are known as the standard tolerances. (IT stands for ISO Tolerance series, and ISO for the International Organisation for Standardization.)

In engineering drawing, it is necessary to provide the nominal size tolerance.

(a) If the function or the economical aspect of the manufacturing process of the workpieces requires that certain limiting dimensions be maintained;

(b) If the parts are required to have a fit;

(c) If the parts are separately finished and need to be assembled without post-machining;

(d) If the parts are to be interchangeable, e. g. spare parts;

(e) If it is necessary that the part be toleranced so that it can be held in a particular fixture for finishing operations.

In ordinary cases, free measure tolerances should suffice.

The tolerance should be so chosen that it just serves the purpose of the relevant application of the workpieces for which it is meant and also ensures interchangeability. The finer the tolerance, the more is the cost of production.

Choice of Tolerance Grades

Grades IT01 to IT7 are used mainly for gauges. The grades 01 to 4 are achieved through lapping, honing, and the finest grinding.

Grades IT5 to IT11 are primarily for the fit of workpieces and are achieved through metal-cutting machining processes, such as fine turning, milling, shaping, planing, grinding, and reaming.

Grades IT12 to IT18 are found suitable for rougher production tolerances in processes not involving metal-cutting, such as forging, rolling, casting, pressing, and drawing.

New Words and Phrases

1. terminology [təːmiˈnɔlədʒi] n. 术语，词汇
2. mating [ˈmætiŋ] n. & adj. 配合（的），相连（的）
3. permissible [pəˈmisəbl] adj. 可允许的，许可的
4. algebraic [ˌældʒiˈbreik] adj. 代数的
5. forego [fɔːˈgəu] v. 先行，在前，居先
6. interference [ˌintəˈfiərəns] n. 过盈
7. finished [ˈfiniʃt] adj. 完美的，精加工的，完工的
8. pictorially [pikˈtɔːriəli] adv. 用（插）图地，如绘成图画地
9. depict [diˈpikt] vt. 画，叙述

10. designate ['dezigneit] *vt.* 标出，把……定义为
11. remoteness [ri'məutnis] *n.* 偏（疏）远
12. closeness [kləuznis] *n.* 接近（程度）
13. subdivide [ˌsʌbdi'vaid] *v.* 细（区）分，再（划）分
14. empirical [em'pirikəl] *adj.* 经验（主义）的
15. micrometer [maikrɔ'mitə] *n.* 微米，千分尺
16. post-machining [pəustmə'ʃiːniŋ] *n.* 后续（期）加工
17. fixture ['fikstʃə] *n.* 夹具，夹紧装置
18. suffix ['sʌfiks] *v.* 满足（……的需要）
19. lapping ['læpiŋ] *n.* 研磨，抛（磨）光
20. honing ['həuniŋ] *n.* 珩（搪）磨
21. grinding ['graindiŋ] *n.* 磨削
22. shaping ['ʃeipiŋ] *n.* 成形加工
23. planning/planing ['plæniŋ] *n.* 刨（削，平）
24. ream [riːm] *vt.* 铰孔，铰大……的口径
25. shaft-basis system   基轴制
26. hole-basis system   基孔制
27. grade (quality) of tolerance   公差等级

Notes

The tolerance should be so chosen that it just serves the purpose of the relevant application of the workpieces for which it is meant and also ensures interchangeability.

公差选择时，应使工件满足相关应用场合的需要，并能保证互换性的要求。

so…that 引导的句子为结果状语从句。

for which 引导的句子为定语从句，修饰先行词 the workpieces。

Exercises

1. Translate the following words into Chinese.

   a) the nominal size or nominal dimension

   b) actual size and limits of size

   c) upper deviation and lower deviation

   d) clearance fit and interference fit

   e) International Organization for Standardization

2. Answer the following questions.

   a) What is tolerance?

   b) Why is tolerance important in manufacturing?

   c) How do you indicate the tolerances on drawing?

   d) Why is the hole-basis system used extensively in modern engineering design?

   e) How many types of fit are there in this text? Describe three main types of fit in detail.

Unit 3　Heat Treatment of Metal

Heat treatment is the operation of heating and cooling a metal in its solid state to change its physical properties. According to the procedure used, steel can be hardened to resist cutting action and abrasion, or it can be softened to permit machining. With the proper heat treatment, internal stresses may be removed, grain size reduced, toughness increased, or a hard surface produced on a ductile interior.

The following discussion applies principally to the heat treatment of plain-corban steels. With this process, rate of cooling is the controlling factor, rapid cooling from above the critical range results in hard structure, whereas very slow cooling produces the opposite effect.

Hardening: In any heat-treating operation, the rate of heating is important. Heat flows from the exterior to the interior of steel at a definite maximum rate. If steel is heated too fast, the outside of the part becomes hotter than the interior. A uniform structure is hard to obtain.

The hardness that can be obtained from a given treatment depends upon the following three factors:

1. Quenching rate;
2. Carbon content;
3. Workpiece size.

Rapid quenching is needed to harden low carbon and medium plain carbon steels. Water is generally used as a quench for these steels. For high-carbon or alloy steel, oil is used. Its action is not as severe as that of water. Where extreme cooling is desired, brine is used.

The maximum degree of hardness obtainable in steel by direct hardening is determined largely by the carbon content. Steel with a low carbon content will not respond greatly to the hardening process. Carbon steels are generally considered as shallow hardening steels. The hardening temperature varies for different steels. The temperature depends upon the carbon content.

The temperature at which steel is usually quenched for hardening is known as the hardening temperature. It is usually 10℃ to 38℃ above the upper critical temperature at which structural change takes place.

Tempering: Hardening makes high-carbon steels and tool steels extremely hard and brittle and not suitable for most uses. By tempering or "drawing", internal stresses developed by the hardening process are relieved. Tempering increased the toughness of the hardened piece. It also seems to make them more plastic or ductile.

Annealing: Annealing consists of heating steel slightly above its critical range and cooling very slowly. Annealing relieves internal stresses and strain caused by previous heat treatment, machining, or other cold-working processes. The type of steel governs the temperature to which the steel is heated for the annealing process. The purpose for which annealing is being done also governs the annealing temperature.

There are three different types of annealing processes used in industry: (1) full annealing,

(2) process annealing, (3) spheroidizing.

Full annealing is used to produce maximum softness in steel. Machinability is improved. Internal stresses are relieved. Process annealing is also called stress relieving. It is used for relieving internal stresses that have occurred during cold-working or machining processes. Spheroidizing is used to produce a special kind of grain structure that is relatively soft and machinable. This process is generally used to improve the machinability in high-carbon steels and in wire-drawing processes.

Normalizing: Normalizing is a process used to relieve the internal stresses due to hot-working, cold-working, and machining. The process consists of heating steel slightly above the upper critical range 30℃ to 50℃ and cooling to room temperature after holding for a while. This process is usually used with low and medium-carbon as well as alloy steels. Normalizing removes all previous effects due to heat treatment.

New Words and Phrases

1. anneal [ə'ni:l] n. 退火
2. heat treatment 热处理
3. quenching [kwentʃiŋ] n. 淬火
4. temper ['tempə] n. 回火
5. brine [brain] n. 盐水
6. medium ['mi:diəm] adj. 中等的，中间的; n. 介质
7. ductile ['dʌktail] adj. 柔软的，易延展的
8. brittle ['britl] adj. 易碎的
9. process annealing 低温退火；中间退火
10. full annealing 完全退火
11. machinability [məˌʃi:nəbiliti] n. 切削性，机械加工性
12. spheroidizing ['sfiərɔidaiziŋ] n. 球化退火
13. normalizing ['nɔ:məˌlaiziŋ] n. 正火
14. stress relieving 去应力退火

Notes

1. The maximum degree of hardness obtainable in steel by direct hardening is determined largely by the carbon content.

说明：obtainable in steel by direct hardening 做定语修饰 hardness，主语为 The maximum degree of hardness，谓语为 is determined。

2. The temperature at which steel is usually quenched for hardening is known as the hardening temperature.

通常使钢淬火变硬的温度，称为淬火温度。

句中"at which steel is usually quenched for hardening"为定语从句，修饰"The temperature"。

is known as：被称为……。

Exercises

1. Fill in the blanks with the information given in the text.

   a) _____ relieves internal stresses and strain, there are _____ types of it used in industry.

   b) Tempering increased the _____ of the hardened piece. It also seems to make them more _____ or _____.

2. Translate the following words into Chinese.

   a) heat treatment; b) quenching rate; c) process annealing; d) spheroidizing; e) stress relieving

3. Answer the following question according to the text above.

   a) How many methods of heat treatment of steel are there in industry? Please try to describe them using the words in this passage or of your own in detail.

   b) What does tempering mean?

   c) What does normalizing mean?

Unit 4 Forming

Forming can be defined as a process in which the desired size and shape are obtained through the plastic deformations of a material. The stresses induced during the process are greater than the yield strength, but less than the fracture strength of the material. The type of loading may be tensile, compressive, bending, shearing, or a combination of these. This is a very economical process as the desired shape, size, and finish can be obtained without any significant loss of the material. Moreover, a part of the input energy is fruitfully utilized in improving the strength of the product through strain hardening.

The forming processes can be grouped under two broad categories, namely, cold forming and hot forming. If the working temperature is higher than the recrystallization temperature of the material, then the process is called hot forming. Otherwise the process is termed as cold forming. The flow stress behavior of a material is entirely different above and below its recrystallization temperature. During hot working, a large amount of plastic deformation can be imparted without significant strain hardening. This is important because a large amount of strain hardening renders the material brittle. The frictional characteristics of the two forming processes are also entirely different. For example, the coefficient of friction in cold forming is generally of the order of 0.1, whereas that in hot forming can be as high as 0.6. Furthermore, hot forming lowers down the material strength so that a machine with a reasonable capacity can be used even for a product having large dimensions.

The typical forming processes are rolling, forging, drawing, deep drawing, bending, and extrusion. For a better understanding of the mechanics of various forming operations, we shall briefly discuss each of these processes.

Rolling

In this process, the job is drawn by means of friction through a regulated opening between two power-driven rolls. The shape and size of the product are decided by the gap between the rolls and their contours. This is a very useful process for the production of sheet metal and various common sections, e.g., rail, channel, angle, and round.

Forging

In forging, the material is squeezed between two or more dies to alter its shape and size. Depending on the situation, the dies may be open or closed.

Drawing

In this process, the cross-section of a wire or that of a bar or tube is reduced by pulling the workpiece through the conical orifice of a die. When high reduction is required, it may be necessary to perform the operation in several passes.

Deep Drawing

In deep drawing, a cup-shaped product is obtained from a flat sheet metal with the help of a punch and a die. The sheet metal is held over the die by means of a blank holder to avoid defects in the product.

Bending

As the name implies, this is a process of bending a metal sheet plastically to obtain the desired shape. This is achieved by a set of suitably designed punch and die.

Extrusion

This is a process basically similar to the closed die forging. But in this operation, the workpiece is compressed in a closed space, forcing the material to flow out through a suitable opening, called a die. In this process, only the shapes with constant cross-sections (die outlet cross-section) can be produced.

Advantages and Disadvantages of Hot and Cold Forming

Now that we have covered the various types of metal working operations, it would only be appropriate that we provide an overall evaluation of the hot and cold working processes. Such a discussion will help in choosing the proper working conditions for a given situation.

During hot working, a proper control of the grain size is possible since active grain growth takes place in the range of the working temperature. As a result, there is no strain hardening, and therefore there is no need of expensive and time-consuming intermediate annealing. Of course, strain hardening is advisable during some operations (viz., drawing) to achieve an improved strength; in such cases, hot working is less advantageous. Apart from this, strain hardening may be essential for a successful completion of some processes (e.g., in deep drawing, strain hardening prevents the rupture of the material around the bottom circumference where the stress is maximum). Large products and high strength materials can be worked upon under hot conditions, since the elevated temperature lowers down the strength and consequently the work load. Moreover, for most materials, the ductility increases with temperature and, as a result, brittle materials can also be worked upon by the hot working operation. It should, however, be remembered that there are certain materials (viz., steels containing sulphur) which become more brittle at elevated temperatures. When a very accurate dimensional control is required, hot working is not advised because of shrinkage and loss of surface metal due to scaling. Moreover, surface finish is poor due to oxide formation and scaling.

The major advantages of cold working are that it is economical, quicker, and easier to handle, because here no extra arrangements for heating and handling are necessary. Furthermore, the mechanical properties normally get improved during the process due to strain hardening. What is more, the control of grain flow directions adds to the strength characteristics of the product. However, apart from other limitations of cold working (viz., difficulty with high strength, brittle materials, and large product size), the inability of the process to prevent the significant reduction in corrosion resistance is an undesirable feature.

New Words and Phrases

1. fracture ['fræktʃə] n. 断裂
2. strain hardening 应变硬化，加工硬化，冷作硬化
3. cold forming 冷成形，冷态成形，冷作成形
4. hot forming 热成形
5. deep drawing 深拉，深冲（压）
6. gap [gæp] n. 缺口，裂口，间隙，缝隙，差距
7. channel ['tʃænl] n. 槽钢
8. superstructure ['sjuːpəˌstrʌktʃə] n. 上部结构
9. conical ['kɔnikəl] adj. 圆锥的，圆锥形的
10. orifice ['ɔːrəfis] n. 孔，节流孔
11. shrinkage ['ʃriŋkidʒ] n. 收缩
12. inability [ˌinə'biliti] n. 无能，无力

Notes

1. Furthermore, hot forming lowers down the material strength so that a machine with a reasonable capacity can be used even for a product having large dimensions.
1) so that 引导一个结果状语从句，表示"以至于……"。
2) having large dimensions 为分词短语作定语，修饰 product。
2. It should, however, be remembered that there are certain materials (viz., steels containing sulphur) which become more brittle at elevated temperatures.
1) 本句中"It"为形式主语，其真正的主语为"that"引导的从句。
2) which 引导的从句为定语从句，修饰先行词"materials"。

Exercises

1. Match the following terms in Column (A) with the descriptions in Column (B).
　　(A)　　　　　　　　　　(B)
(　　) 1. drawing　　　　a) The shape and size of the product are decided by the gap

(   ) 2. forging   b) The material is squeezed between two or more dies to alter its shape and size.

(   ) 3. rolling   c) The cross-section of a wire or that of a bar or tube is reduced by pulling the workpiece through the conical orifice of a die.

(   ) 4. extrusion   d) A cup-shaped product is obtained from a flat sheet metal with the help of a punch and a die.

(   ) 5. bending   e) A desired shape can be obtained by bending a metal sheet plastically.

(   ) 6. deep drawing   f) Only the shapes with constant cross-sections can be produced.

(   ) 7. cold forming   g) The desired size and shape are obtained through the plastic deformation of a material.

(   ) 8. forming   h) The working temperature is lower than the recrystallization temperature of a material.

(   ) 9. die   i) The working temperature is higher than the recrystallization temperature of a material.

(   ) 10. hot forming   j) It is a suitable opening through which the material is formed to flow out.

2. Answer the following questions in detail according to the text.

a) How do you define the term "forming" in English according to the text?

b) What technical features should we consider when carrying out various forming processes mentioned in the text?

Reading Material A

Kinds of Steel

There are two general kinds of steels: carbon steel and alloy steel. Carbon steel contains only iron and carbon, while alloy steel contains some other "alloying elements" such as nickel, chromium, manganese, molybdenum, tungsten, vanadium, etc.

1. Carbon steels

(1) Low carbon steel contains from 0.05 to 0.15 percent carbon, this steel is also known as machine steel.

(2) Medium carbon steel contains from 0.15 to 0.60 percent carbon.

(3) High carbon steel contains from 0.60 to 1.50 percent carbon, this steel is sometimes called "tool steel".

2. Alloy steels

(1) Special alloy steel, such as nickel steel, chromium steel.

(2) High-speed steel also known as self-hardening steel.

The properties of carbon steels depend only on the percentage of carbon they contain.

Low carbon steels are very soft and can be used for bolts and for machine parts that do not need strength.

Medium carbon steel is a better grade and stronger than low carbon steel. It is also more difficult to cut than low carbon steel.

Heating it to a certain temperature and then quickly cooling in water may harden high carbon steel. The more carbon the steel contains and the quicker the cooling is, the harder it becomes. Because of its high strength and hardness, this grade of steel may be used for tools and working parts of machines. But for some special uses, for example, for gears, bearings, springs, shafts and wire, carbon steels cannot be always used because they have no properties needed for these parts.

Some special alloy steels should be used for such parts because the alloying elements make them tougher, stronger, or harder than carbon steels. Some alloying elements cause steel to resist corrosion, and such steels are called stainless steels.

Heat-resistant steel is made by adding some tungsten and molybdenum, while manganese increases the wear resistance of steel. Vanadium steels resist corrosion and can stand shocks and vibration.

Tools made of high-speed steel containing tungsten, chromium, vanadium, and carbon, may do the work at much higher speeds than carbon tool steels.

New Words and Phrases

1. molybdenum [məˈlibdinəm] *n.* 钼
2. stainless [ˈsteinlis] *adj.* 不锈的
3. vibration [vaiˈbreiʃən] *n.* 振动，颤动
4. heat-resistant steel   耐热钢
5. self-hardening steel   自硬钢
6. high-speed steel   高速钢

Reading Material B

Dimension and Tolerance

In dimensioning a drawing, the numbers placed in the dimension lines represent dimensions that are only approximate and do not represent any degrees of accuracy unless so stated by the designer. The numbers are termed as nominal size. The nominal size of a component dimension is arrived at as a convenient size based on the design process. However, it is almost impossible to produce any component to the exact dimension through any of the known manufacturing processes. Even if a component is perceived to be made to the exact dimension by manual processes, the actual measurement with a high resolution measuring device will show that this is an incorrect perception. It is therefore customary in engineering practice to allow a permissible deviation from the nominal size, which is termed as tolerance. Tolerance on a dimension can also specify the degree of accuracy. For example, a shaft might have a nominal size of 63.5mm, if a variation of ±0.08mm could be permitted, the dimension would be stated 63.5±0.08mm.

In engineering, when a product is designed, it consists of a number of parts and these parts mate with each other in some form. In the assembly, it is important to consider the type of mating or fit between two parts which will actually define the way the parts are to behave during the working of the assembly. Take for example a shaft and hole, which will have to fit together. In the simplest case, if the dimension of the shaft is lower than the dimension of the hole, then there will be clearance. Such a fit is termed clearance fit. Alternatively, if the dimension of the shaft is more than that of the hole, then it is termed interference fit.

Dimensions given close tolerances mean that the part must fit properly with some other parts. Both must be given tolerances in keeping with the allowance desired, the manufacturing processes available, and the minimum cost of production and assembly that will maximize profit. Generally speaking, the cost of a part goes up as the tolerance is decreased. If a part has several or more surfaces to be machined, the cost can be excessive when little deviation is allowed from the nominal size.

Allowance, which is sometimes confused with tolerance, has an altogether different meaning. It is the minimum clearance space intended between mating parts and represents the condition of the tightest permissible fit. If a shaft, size $1.498_{-0.003}^{-0.000}$, is to fit a hole of size $1.500_{-0.000}^{+0.003}$, the minimum size hole is 1.500 and the maximum size shaft is 1.498. Thus the allowance is 0.002 and the maximum clearance is 0.008 as based on the minimum shaft size and maximum hole dimension.

Tolerances may be either unilateral or bilateral. Unilateral tolerance means that any variation is made in only one direction from the nominal or basic dimension. Referring to the previous example, the hole is dimensioned $1.500_{-0.000}^{+0.003}$, which represents a unilateral tolerance. Here the nominal size 1.500 is allowed to vary between 1.503 and 1.5. If the dimensions were given as 1.500±0.003, the tolerance would be bilateral; that is, it would vary both over and under the nominal dimension. In bilateral tolerance, the variation of the limits can be uniformed as 30.00±0.02. The dimension varies from 30.02 to 29.98. Alternatively the allowed deviation can be different as $30.00_{-0.10}^{+0.05}$. Here the dimension varies from 30.05 to 29.90. Sometimes, the nominal size may be outside the allowable limits. For example, a given dimension is to vary from 29.95 to 29.85. It can be written as $29.95_{-0.10}^{+0}$ or $30.00_{-0.15}^{-0.05}$. The second form is preferred since it contains the nominal size 30. The unilateral system permits changing the tolerance while still retaining the same allowance or type of fit. With the bilateral system, this is not possible without changing the nominal size dimension of one or both of the two mating parts. In mass production, where mating parts must be interchangeable, unilateral tolerances are customary. To have an interference or force fit between mating parts, the tolerances must be such as to create a zero or negative allowance.

New Words and Phrases

1. interference fit   干涉配合，静配合

2. close tolerance 紧公差
3. allowance [ə'lauəns] *n.* 容差
4. unilateral [ˌjuːnə'lætərəl] *adj.* 单边的，单向的
5. bilateral [bai'lætərəl] *adj.* 双向的，双边的
6. force fit 压入配合，压紧配合

Reading Material C

Metal Hot Working

As we know, casting is a mechanical working process that forms a molten material into a desired shape by pouring it into a mold and letting it harden. When metal is not cast in a desired manner, it is formed into special shapes by mechanical working processes. Several factors must be considered when determining whether a desired shape is to be cast or formed by mechanical working. If the shape is very complicated, casting will be necessary to avoid expensive machining of mechanically formed parts. On the other hand, if strength and quality of material are the prime factors in a given part, a cast will be unsatisfactory. For this reason, steel castings are seldom used in aircraft work.

There are three basic methods of metal-working. They are hot working, cold working, and extruding. The process chosen for a particular application depends upon the metal involved and the part required, although in some instances you might employ both hot-working and cold-working methods in making a single part.

Almost all steel is hot-worked from the ingot into some form from which it is either hot-or cold-worked to the finished shape. When an ingot is stripped from its mold, its surface is solid, but the interior is still molten. The ingot is then placed in a soaking pit, which retards loss of heat, and the molten interior gradually solidifies. After soaking, the temperature is equalized throughout the ingot, which is then reduced to intermediate size by rolling, making it more readily handled.

Hot-working is the process in which the ingot is deformed mechanically into a desired shape. Hot-working is usually performed at an elevated temperature. At high temperature, scaling and oxidation exist. Scaling and oxidation produce undesirable surface finish. Often times, most ferrous metals need to be cold-worked after hot-working in order to improve the surface finish.

The main principle behind hot-working is to cause plastic deformation within the material. The amount of force needed to perform hot-working is normally less than that for cold-working. As such, the mechanical properties of the material remain unchanged during hot-working. The reason that the properties of the materials are unaltered comes from the fact that the deformation is performed above the metal recrystallization temperature. Plastic deformation occurs with metals when deformed above the recrystallization temperature without any strain hardening. As a matter of fact, the metal usually experiences a decrease in yield strength when hot-worked. Therefore, it is possible to hot-work the metal without causing any fracture.

New Words and Phrases

1. hot-working 热加工
2. cold-working 冷加工
3. soaking pit 均热炉
4. squeeze [skwiːz] v. 挤，压
5. ingot ['iŋgət] n. 工业纯铁
6. scaling ['skeiliŋ] v. 剥落，氧化起皮

Reading Material D

Surface Roughness

As it is not possible to achieve, in practice, a geometrically ideal surface of a workpiece, an engineering drawing must also contain information about the permissible surface conditions of the body. The surface condition is a function of the finishing process adopted. The relevant terms which are now defined are pictorially represented in Fig. ID-1.

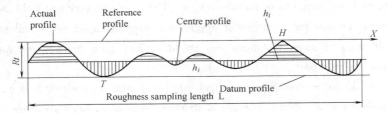

Fig. ID-1 Surface roughness

Actual profile: the profile of the actual surface.

Reference profile: the profile to which the irregularities of surface are referred; it passes through the highest point H of the actual profile.

Datum profile: the profile parallel to the reference profile; it passes through the farthest point T of the actual profile.

Centre profile or mean profile: the profile within the sampling length so placed that the sum of the material-filled areas enclosed above it by the actual profile is equal to the sum of the material-void areas enclosed below it by the actual profile.

Peak-to-valley height Rt: the distance from the datum profile to the reference profile.

Mean roughness index Ra: the arithmetic mean of the absolute values of the distance h_i between the actual and the centre profiles. It is given by

$$Ra = \frac{1}{L}\int_{x=0}^{x=L} |h_i| dx$$

Surface roughness can be represented in several ways in an engineering drawing. The method followed in IS: 696—1972 is given in Fig. ID-2. The basic symbol consists of two legs of unequal length representing the surface under consideration (Fig. ID-2a). If the material by machining is required, a bar is added to the basic symbol (Fig. ID-2b). If the removal of material is not permitted, a circle is added to the basic symbol (Fig. ID-2c). When special surface characteristics have to be indicated, a line is added to the longer leg (Fig. ID-2d). For positions of the specifications of surface roughness and related data with respect to the symbol, see Fig. ID-2e.

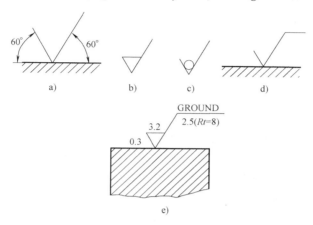

Fig. ID-2 Representation of surface roughness

New Words and Phrases

1. surface roughness　表面粗糙度
2. irregularity [iˈregjuˈlæritiː] n. 不规则（均匀，对称）性
3. datum [ˈdeitəm] n. 基准（点，线，面）

Part II Machine Tools

Unit 5 Lathes

The basic machines that are designed primarily to do turning, facing and boring are called lathes. Very little turning is done on other types of machine tools, and none can do it with equal facility. Because lathe can do boring, facing, drilling, and reaming in addition to turning, their versatility permits several operations to be performed with a single setup of the workpiece.

The essential components of a lathe are depicted in the block diagram of Fig. 5-1. These are the bed, headstock assembly, tailstock assembly, carriage assembly, quick-change gear box, the leadscrew, and feed rod.

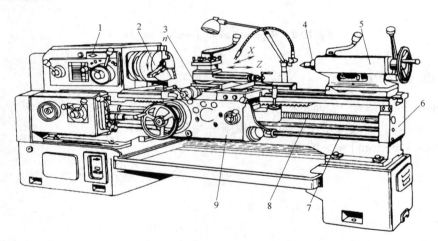

Fig. 5-1 Block diagram of basic components of a lathe
1—headstock 2—chuck 3—tool-post 4—center 5—tailstock
6—bed 7—feed-shaft 8—leadscrew 9—carriage

The bed is the backbone of a lathe. It is usually made of well-normalized or aged gray or nodular cast iron and provides a heavy, rigid frame on which all the other basic components are mounted. Two sets of parallel, longitudinal ways, inner and outer, are contained on the bed, usually on the upper side. Because several other components are mounted and/or move on the ways, they must be made with precision to assure accuracy of alignment. Similarly, proper precaution should be taken in operating a lathe to assure that the ways are not damaged. Any inaccuracy in them usually means that the accuracy of the entire lathe is destroyed. The ways on most modern lathes are surface-hardened to offer greater resistance to wear and abrasion.

The headstock is mounted in a fixed position on the inner ways at one end of the lathe bed. It provides a powered means of rotating the work at various speeds. It essentially consists of a hollow spindle mounted in accurate bearings, and a set of transmission gears similar to a truck transmission through which the spindle can be rotated at a number of speeds.

Because the accuracy of a lathe is greatly dependent on the spindle, it is of heavy construction and mounted in heavy bearings, usually preloaded tapered roller or ball types.

The inner end of the spindle protrudes from the gear box and contains a means for mounting various types of chucks, face plates, and dog plates on it. Whereas small lathes often employ a threaded section to which the chucks are screwed, most large lathes utilize either cam-lock or key-drive taper noses.

The tailstock assembly essentially consists of three parts. A lower casting fits on the inner ways of the bed and can slide longitudinally thereon, with a means for clamping the entire assembly in any desired location. An upper casting fits on the lower one and can be moved transversely upon it on some type of keyed ways. This transverse motion permits aligning the tailstock and headstock spindles and provides a method of turning tapers. The third major component of the assembly is the tailstock quill. This is a hollow steel cylinder, usually about 2 to 3 inches in diameter, that can be moved several inches longitudinally in and out of the upper casting by means of a handwheel and a screw.

The carriage assembly provides the means for mounting and moving cutting tools. The carriage is a relatively flat H-shaped casting that rests and moves on the outer set of ways on the bed.

On most lathes the tool post actually is mounted on a compound rest. This consists of a base, which is mounted on the cross slide so that it can be pivoted about a vertical axis and an upper casting. The upper casting is mounted on ways on this base so that it can he moved back and forth and controlled by means of a short lead screw operated by a handwheel and a calibrated dial.

Manual and powered motion for the carriage, and powered motion for the cross slide, are provided by mechanism within the apron attached to the front of the carriage. Manual movement of the carriage along the bed is effected by turning a handwheel on the front of the apron, which is geared to a pinion on the back side. This pinion engages a rack that is attached beneath the upper front edge of the bed in an inverted position.

To impart powered movement to the carriage and cross slide, a rotating feed rod is provided. The feed rod, which contains a keyway throughout most of its length, passes through the two reversing bevel pinions and is keyed to them.

Modern lathes have a quick-change gear box. The input end of this gear box is driven from the lathe spindle by means of suitable gearing. The output end of the gear box is connected to the feed rod and lead screw. Thus, through this gear train, leading from the spindle to the quick-change gear box, then to the lead screw and feed rod, and then to the carriage, the cutting tool can be made to move a specific distance, either longitudinally or transversely, for each revolution of the spindle.

New Words and Phrases

1. lathe [leið] *n.* 车床

2. leadscrew 丝杠，导（螺）杆
3. headstock ['hedstɔk] n. 主轴箱，床头箱
4. tailstock ['teilstɔk] n. 尾座，尾架
5. carriage ['kæridʒ] n. 拖板，大拖板
6. nodular ['nɔdjulə] adj. 球（团，粒）状的
7. center ['sentə] n. 顶尖
8. alignment [ə'lainmənt] n. 成直线，（直线）对准
9. protrude [prə'truːd] v. （使）伸，突出
10. chuck [tʃʌk] n. 卡盘
11. quill [kwil] n. 活动套筒
12. apron ['eiprən] n. 溜板箱，挡板
13. engage [in'geidʒ] v. 啮合
14. whereas [weər'æz] conj. 考虑到，鉴于
15. headstock assembly 主轴箱组件
16. carriage assembly 溜板箱组件
17. cam-lock 偏心夹，凸轮锁紧
18. face plate 面板，花盘
19. cross slide 中拖板

Notes

1. Because the accuracy of a lathe is greatly dependent on spindle, it is of heavy construction and mounted in heavy bearing, usually preloaded tapered roller or ball types.

因为车床的精度在很大程度上取决于主轴，所以主轴的结构尺寸较大，通常安装在紧密配合的重型圆锥滚子轴承或球轴承中。

（1）it *pron.* 指 spindle。

（2）be of heavy construction 指主语所具有的性质、特征等。

heavy construction 根据上下文需要，意译。

（3）in heavy bearing 中的 heavy 是意译。

（4）usually preloaded tapered roller or hall types 做定语，修饰 heavy bearing。

2. A lower casting fits on the inner ways of the bed and can slide longitudinally thereon, with a means for clamping the entire assembly in any desired location.

底座安装在床身的内侧导轨上，并可以在导轨上作纵向移动，底座上有一个可以使整个尾座组件夹紧在任意位置上的装置。

（1）fit on 安装，穿上

（2）means *n.* 方法，手段，工具

Exercises

1. Translate the following words into Chinese.

   a) turning, facing and boring   b) headstock assembly   c) tailstock assembly

d) carriage assembly   e) lead screw and feed rod

2. Answer the following questions.

a) What types of surfaces can be made by turning?

b) What are the basic machining operations that can be done on a lathe?

c) How many parts does the tailstock assembly consist of? What are they?

d) Describe the function of the lead screw.

e) Tell the essential components of a lathe.

Unit 6   Grinding Classification and Centreless Grinding

Generally, grinding is used as a finishing process to get the desired surface finish, correct size and accurate shape of the product. However, recent researches have shown that grinding can also be used economically for bulk removal of unwanted material just like turning, milling, etc.

Classification of Grinding Processes

Grinding processes may be classified according to the shape of the surface that is ground, type of the grinding machine used or type of the ground product. According to the type of the surface and type of the machine, the following classification is obtained.

1. Surface grinding is used for grinding flat surfaces.

2. Cylindrical grinding is used for grinding external and internal cylindrical surfaces.

3. Centreless grinding is used for the grinding of external and internal cylindrical surfaces. In these processes, machines are different from those used for conventional cylindrical grinding.

4. Form grinding includes the grinding of gears, thread grinding and grinding of splines, etc.

5. The abrasive cut-off process is used for severing metallic and nonmetallic materials with the help of thin grinding wheels rotating at high speed. The process depending on the cutting action of abrasive particles, is assisted by the heat produced by the cutting action.

6. Abrasive belt grinding has acquired an important place in grinding processes. It can be used to grind flat, cylindrical or curved shapes as the belt can be easily made to conform to the shape of the component.

7. Off-hand grinding processes are those in which the workpiece or grinding wheel is held by hand and guided and forced to carry out the grinding processes. Machines like bench grinders, portable grinders, die grinders, etc. are used for the purpose.

However, there are many products for which special grinding machines have been designed, such as crank shaft grinding machines, camshaft grinding machines, etc.

In each category of grinding processes, variations are there according to the type of the product to be ground or according to the type of the controls provided with the machine. Thus for example, in cylindrical grinding, the following different types of machines are manufactured.

1. Plain cylindrical grinding machines.

2. Heavy plain or roll grinding machines.

3. Universal cylindrical grinding machines.

4. CNC cylindrical grinders.
5. Contour grinding machines such as computer controlled cam grinders.
6. Crank shaft grinding machines.

In modern CNC machines, automatic control of workpiece dimensions as well as facility of automatic dressing of the grinding wheel are provided.

Centreless Grinding

In centreless grinding, the job is held neither between centres nor in a chuck as is the case in cylindrical grinding. Instead, the cylindrical workpiece is supported between the grinding wheel, the control wheel and a work rest for external cylindrical surfaces. The workpiece rotation is controlled by the surface speed of the control wheel. The work rest is adjusted so that the workpiece center is kept above the centerline joining the centres of the grinding wheel and the control wheel. The work rest blade is also inclined with respect to the centerline. The work rest blade inclination and the height of workpiece center are important parameters for obtaining precision cylindrical surfaces.

External Centreless Grinding

In order to cope with the different configurations of industrial products, four schemes of external centreless grinding are used in industry.

1. Through-feed centreless grinding.
2. In-feed centreless grinding.
3. End-feed centreless grinding.
4. Combination of in-feed and end-feed centreless grinding.

The through-feed process is meant for plain cylindrical workpieces. The axis of the control wheel is slightly inclined to the axis of the grinding wheel. Thus the workpiece gets two types of motion, i.e. (1) rotation about its axis and (2) linear motion parallel to the axis of the grinding wheel.

The in-feed process is used for stepped cylindrical workpieces which cannot be through-fed. In this case, the control wheel is retracted, the workpiece is fed and the control wheel is then advanced to carry out the grinding process.

Internal Centreless Grinding

For internal cylindrical grinding, two schemes are used. In the first scheme, the tubular workpiece is supported between the control wheel, the support wheel and a pressure roll, such that centers of the control wheel, workpiece and the grinding wheel all lie on the same line. This is called on-centre internal centreless grinding.

In the second scheme, the workpiece is also supported between the control wheel, the support wheel and a pressure roll, but in this case the grinding wheel centre does not lie on the line joining the control wheel centre and workpiece centre. This is called off-centre internal centreless grinding.

In the first scheme, the wall thickness of the tubular workpiece is accurately ground even for very thin tubes.

New Words and Phrases

1. bulk [bʌlk] *n.* 整体，容量
2. lap [læp] *n. & v.* 研磨，抛光
3. hone [həun] *n. & v.* 磨
4. spline [splain] *n.* 花键（轴）
5. belt [belt] *n.* 层，界
6. parameter [pə'ræmitə] *n.* 参数，系数
7. stepped [stept] *adj.* 有阶梯的，成梯状的
8. configuration [kən,figju'reiʃən] *n.* 外形，轮廓
9. scheme [ski:m] *n.* 计划，方案
10. index ['indeks] *vt.* 分度
11. through-feed 纵向进给
12. in-feed 横向进给
13. end-feed 纵向定程进给

Notes

1. Off-hand grinding processes are those in which the workpiece or grinding wheel is held by hand and guided and forced to carry out the grinding process.

手工磨削加工：在这一加工过程中，手持工件或砂轮移动并加工。

be held... and guided and forced 是三个并列的谓语。

2. The tubular workpiece is supported between the control wheel, the support wheel and a pressure roll, such that centers of the control wheel, workpiece and the grinding wheel all lie on the same line.

管状工件放置在导轮、支撑轮和加压辊之间，导轮、工件和砂轮的中心全都在同一条直线上。

Exercises

1. Translate the following words into Chinese.

a) In-feed centreless grinding   b) thread grinding   c) through-feed centreless grinding

d) the axis of the grinding wheel e) End-feed centreless grinding

f) universal cylindrical grinding machines   g) CNC cylindrical grinding machines

h) crank shaft grinding machines

2. Answer the following questions.

a) Why is grinding generally used as a finishing process?

b) Introduce the classification of grinding processes briefly.

c) Which types of grinding can be used for grinding external and internal cylindrical surfaces?

d) What is through-feed centreless grinding?

e) Which types of methods can be used for grinding of gears?

Unit 7 Milling

Milling is a basic machining process in which the surface is generated by the progressive formation and removal of chips of material from the workpiece as it is fed to a rotating cutter in a direction perpendicular to the axis of the cutter. In some cases, the workpiece is stationary and the cutter is fed to the work. In most instances, a multiple-tooth cutter is used so that the metal removal rate is high, and frequently the desired surface is obtained in a single pass of the work.

The tool used in milling is known as a milling cutter. It usually consists of a cylindrical body which rotates on its axis and contains equally spaced peripheral teeth that intermittently engage and cut the workpiece. In some cases, the teeth extend part way across one or both ends of the cylinder.

Because the milling principle provides rapid metal removal and can produce good surface finish, it is particularly well-suited for mass-production work, and excellent milling machines have been developed for this purpose. However, very accurate and versatile milling machines of a general-purpose nature also have been developed that are widely used in job-shop and tool and die work. A shop that is equipped with a milling machine and an engine lathe can machine almost any type of product with a suitable size.

Types of Milling Operations

Milling operations can be classified into two broad categories, each of which has several variations.

1. In peripheral milling, a surface is generated by teeth located in the periphery of the cutter body; the surface is parallel with the axis of rotation of the cutter. Both flat and formed surfaces can be produced by this method. The cross section of the resulting surface corresponds to the axial contour of the cutter. This procedure is also called slab milling.

2. In face milling, the generated flat surface is at right angles to the cutter axis and is the combined result of the actions of the portions of the teeth located on both the periphery and the face of the cutter. The major portion of the cutting is done by the peripheral portions of the teeth with the face portions providing a finishing action.

The basic concepts of peripheral milling and face milling are illustrated in Fig. 7-1. Peripheral milling operations usually are performed on machines having horizontal spindles, whereas face milling is done on both horizontal and vertical spindle machines.

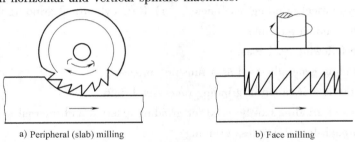

a) Peripheral (slab) milling   b) Face milling

Fig. 7-1 Tool-work relationships in peripheral and face milling

Milling Cutters

Milling cutters can be classified in several ways. One method is to group them into two broad classes based on tooth relief as follows.

1. Profile-cutters have relief provided on each tooth by grinding a small land back of the cutting edge. The cutting edge may be straight or curved.

2. In form or cam-relieved cutters, the cross section of each tooth is an eccentric curve behind the cutting edge, thus providing relief. All sections of the eccentric relief, parallel with the cutting edge, must have the same contour as the cutting edge. Cutters of this type are sharpened by grinding only the face of the teeth, with the contour of the cutting edge thus remaining unchanged.

Another useful method of classification is according to the method of mounting the cutter. Arbor cutters are those that have a center hole so they can be mounted on an arbor. Shank cutters have either tapered or straight integral shank. Those with tapered shanks can be mounted directly in the milling machine spindle, whereas straight-shank cutters are held in a chuck. Facing cutters usually are bolted to the end of a stub arbor.

Types of Milling Cutters

Plain milling cutters are cylindrical or disk-shaped, having straight or helical teeth on the periphery. They are used for milling flat surfaces. This type of operation is called plain or slab milling. Each tooth in a helical cutter engages the work gradually, and usually more than one tooth cuts at a given time. This reduces shook chattering tendencies and promotes a smoother surface. Consequently, this type of cutter usually is preferred over one with straight teeth.

Side milling cutters are similar to plain milling cutters except that the teeth extend radially part way across one or both ends of the cylinder toward the center. The teeth may be either straight or helical. Frequently these cutters are relatively narrow, being disklike in shape. Two or more side milling cutters often are spaced on an arbor to make simultaneous, parallel cuts, in an operation called straddle milling.

Interlocking slotting cutters consist of two cutters similar to side mills, but made to operate as a unit for milling slots. The two cutters are adjusted to the desired width by inserting shims between them.

Staggered-tooth milling cutters are narrow cylindrical cutters having staggered teeth, and with alternate teeth having opposite helix angles. They are ground to cut only on the periphery, but each tooth also has chip clearance ground on the protruding side. These cutters have a free cutting action that makes them particularly effective in milling deep slots.

Metal-slitting saws are thin, plain milling cutters, usually from 1/32 to 3/16 inch thick, which have their sides slightly "dished" to provide clearance and prevent binding. They usually have more teeth per inch of diameter than ordinary plain milling cutters and are used for milling deep, narrow slots and for culling-off operations.

New Words and Phrases

1. peripheral [pəˈrifəiəl] *adj.* 圆周的、周边的,外部的

2. contour ['kɔn'tuə] n. 轮廓（线），外形
3. slab [slæb] n. 平面，平板
4. land [lænd] n. 刀棱面，齿刃
5. eccentric [ik'sentrik] adj. 偏心的，不同圆心的
6. integral ['intigrəl] adj. 整（数，体）的，完整的
7. staggered ['stægəd] adj. 交错的
8. slot [slɔt] n. 槽（沟），（裂）缝
9. straddle ['strædl] adj. 跨式的
10. interlocking ['intə'lɔkiŋ] adj. 可联动的，互锁的
11. shim [ʃim] n. （薄）垫
12. job-shop 机修车间
13. be at right angles to 与……成直角

Notes

1. It usually consists of a cylindrical body which rotates on its axis and contains equally spaced peripheral teeth that intermittently engage and cut the workpiece.

它通常是一个绕其轴线旋转并且周边带有同间距齿的圆柱体，铣刀齿间歇性接触并切削工件。

（1）which rotates on its axis and contains equally spaced peripheral teeth 为定语从句，修饰 acylindrical body。在该定语从句中 contains 与 consists of 为两个并列谓语。

（2）that intermittently engage and cut the workpiece 为定语从句，修饰前置词 teeth。

2. In face milling, the generated flat surface is at right angles to the cutter axis and is the combined result of the action of the portions of the teeth located on both the periphery and the face of the cutter.

端面铣削时，铣削平面与刀具的轴线垂直，被加工平面是位于刀具周边和端面上的齿综合作用形成的。

（1）该句为"主语+系动词+表语"结构，主语为 the generated flat surface。

is at right angles to the cutter axis 和 is the combined result of the action of the portions of the teeth located on both the periphery and the face of the cutter 为两个"系动词+表语"结构。

（2）过去分词短语 located on both the periphery and the face of the cutter 做后置定语，修饰前置词 the teeth。

（3）that the cutting force tends to hold the work against the machine table, permitting lower clamping forces to be employed 为表语从句。

（4）permitting lower clamping forces to be employed 为分词短语，在表语从句中做状语。

Exercises

1. Translate the following words into Chinese.

a）a multiple-tooth cutter   b）the metal removal rate   c）produce good surface finish

d）slab milling   e）face milling   f）up milling   g）down milling

h) arbor cutters and shank cutters
2. Answer the following questions.
a) What is milling?
b) What is peripheral milling?
c) What is face milling?
d) How many types of milling cutters are there? What are they?
e) Why are helical teeth often used for milling cutters?

Unit 8   Drills and Drill Presses

Drills are classified by several features: (1) drill diameter, from 0.30 to 100mm, (2) the steel in the drill, either carbon or high-speed steel, (3) the type of shank, either parallel or tapered, and (4) the length of the drill.

Parallel-shank drills can be used only in chucks and are dependent for alignment on the condition of both chuck and drill shank. Taper shank drills are more satisfactory in this respect. At the end of the taper shank, there is a small flat tang which engages with a slot in the spindle. The taper is the Standard Morse Taper, in sizes ranging from No. 1, the smallest, to No. 6. Where the drill taper shank is smaller than the taper in the spindle, the two can be matched by fitting a sleeve to the drill shank so that a drill with a No. 1 Morse taper, for example, could be fitted into a spindle made with a No. 2 Morse taper.

The web thickness increases from the drill tip to the run-out of the flutes to give extra strength, but this has the effect of increasing the length of the chisel edge as the drill is worn away by regrinding. This longer edge calls for extra feed pressure to overcome the resistance offered by this part of the drill, but this trouble can be rectified by reducing the web thickness by grinding on a thin, round-edged wheel.

Twist drills should be reground immediately when there is any sign of inefficient working which will be revealed by: (1) the need for excessive feed pressure to make the drill cut, (2) the ejection of scored cutting which indicates chipped cutting-edges, and (3) chattering or screaming from the drill when pressure is applied. This is caused by the drill rubbing instead of cutting and will quickly cause overheating.

There are three basic types of drill presses used for general drilling operations: the sensitive drill press, the upright drilling machine, and the radial arm drill press. The sensitive drill press, as the name implies, allows the operator to "feel" the cutting action of the drill as his hand feeds it into the work. These machines are either bench or floor mounted. Since these drill presses are used for light duty applications only, they usually have a maximum drill size of 1/2 in diameter.

The sensitive drill press has four major parts, not including the motor: the head, column, table, and base. The spindle rotates within the quill, which does not rotate but carries the spindle up and down. The spindle shaft is driven by a stepped-vee pulley and belt or by a variable speed drive.

The upright drill press is very similar to the sensitive drill press, but it is made for much heavier work. The drive is more powerful and many types are gear driven, so they are capable of drilling holes to two inches or more in diameter. The motor must be stopped when changing speeds on a gear drive of the drill press. If it doesn't shift into the selected gear, turn the spindle by hand until it meshes. Since power feeds are needed to drill these large size holes, these machines are equipped with power feed mechanisms that can be adjusted by the operator. The operator may either feed manually with a lever or a hand wheel or he may engage the power feed. A mechanism is provided to raise and lower the table.

The radial arm drill press is the most versatile drilling machine. Its size is determined by the diameter of the column and the length of the arm measured from the center of the spindle to the outer edge of the column. It is useful for operations on large castings that are too heavy to be repositioned by the operator for drilling each hole. The work is clamped to the table or base, and the drill can then be positioned where it is needed by swinging the arm and moving the head along the arm. The arm and the head can be raised or lowered on the column and then locked in place. The radial arm drill press is used for drilling small to very large holes and for boring, reaming, counterboring, and countersinking. Like the upright machine, the radial arm drill press has a power feed mechanism and a hand feed lever.

New Words and Phrases

1. dress press　钻床
2. web ［web］ n. 钻芯
3. flute ［flu:t］ n. 凹槽, 出屑槽
4. chisel ［'tʃizəl］ n. 钻头横刃
5. rectify ［'rektifai］ vt. 消除, 改正
6. twist drill　麻花钻头
7. ejection ［i'dʒekʃən］ n. 挤出, 抛出
8. score ［skɔ:］ v. 刻划, 研刻
9. chatter ［'tʃætə］ v. 振动
10. sensitive drill press　台式钻床, 高速手压钻床
11. upright drill press (upright drilling machine)　立式钻床
12. radial aim drill press　摇臂钻床
13. stepped-vee pulley　台阶式 V 形槽带轮
14. reposition ［,ri:pə'ziʃən］ vt. 改变……的位置

Notes

1. The taper is the Standard Morse Taper, in sizes ranging from No. 1, the smallest, to No. 6. Where the drill taper shank is smaller than the taper in the spindle, the two can be matched by fitting a sleeve to the drill shank so that a drill with a No. 1 Morse taper, for example, could be fitted into a spindle made with a No. 2 Morse taper.

1) "the Standard Morse Taper" 指 "莫氏标准锥度"。

2）句中 Where 引导的定语从句，修饰 No. 1 to No. 6。

3）so that 引导的目的状语从句，表示"以便……"。

2. If it doesn't shift into the selected gear, turn the spindle by hand until it meshes. Since power feeds are needed to drill these large size holes, these machines are equipped with power feed mechanisms that can be adjusted by the operator.

1）"If it doesn't shift into the selected gear"为条件状语从句，表示"如果……"。

2）"Since power feeds are needed to drill..."为 Since 引导的原因状语从句，表示"由于……"。

Exercises

1. Translate the following words into Chinese.
   a) sensitive drill press   b) upright drill press   c) radial aim drill press
   d) parallel-shank drill   e) taper shank drill   f) Standard Morse Taper
2. Answer the following questions.
   a) When should the twist drills be reground?
   b) How many basic types of drill presses are used for general drilling operations?
   c) How many major parts does the sensitive drill press has?
   d) What is the radial arm drill press used for?

Reading Material A

Spur and Helical Gears

A gear having tooth elements that are straight and parallel to its axis is known as a spur gear. A spur pair can be used to connect parallel shafts only. Parallel shafts, however, can also be connected with gears of another type, and a spur gear can be mated with a gear of a different type.

To prevent jamming as a result of thermal expansion, to aid lubrication, and to compensate for unavoidable inaccuracies in manufacture, all power-transmitting gears must have backlash. This means that on the pitch circles of a mating pair, the space width on the pinion must be slightly greater than the tooth thickness on the gear, and vice versa. On instrument gears, using a gear split down its middle, one half being ratable relative to the other can eliminate backlash. A spring forces the split gear teeth to occupy the full width of the pinion space.

Helical gears have certain advantages, for example, when connecting parallel shafts, they have a higher load carrying capacity than spur gears with the same tooth numbers and cut with the same cutter. Because of the overlapping action of the teeth, they are smoother in action and can operate at higher pitch-line velocities than spur gears. The pitch-line velocity is the velocity in the pitch circle. Since the teeth are inclined to the axis of rotation, helical gears create an axial thrust. If used single, this thrust must be absorbed in the shaft bearings. The thrust problem can be overcome by cutting two sets of opposed helical teeth on the same blank. Depending on the method of manufacture, the gear may be of the continuous-tooth herringbone variety or a double-helical gear with a space between the two halves to permit the cutting tool to run out. Double-helical gears are well suited for the efficient transmission of power at high speeds.

Helical gears can also be used to connect nonparallel, non-intersecting shafts at any angle to one another. Ninety degrees is the commonest angle, at which such gears are used.

New Words and Phrases

1. spur ［spəː］ *n.* ［建］凸壁；支撑物
2. helical ［ˈhelikəl］ *adj.* 螺旋形的；螺旋线的
3. spur gear　正齿轮
4. helical gear　斜齿轮
5. thrust ［θrʌst］ *v.* ［机］推力；侧向压力；插；猛推
6. lubrication ［ˌluːbriˈkeiʃən］ *n.* 润滑
7. compensate ［ˈkɔmpenseit］ *v.* 偿还，补偿，付报酬
8. backlash ［ˈbæklæʃ］ *n.* 轮齿隙；反斜线（\）；后座；后冲
9. ratable ［ˈreitəbl］ *adj.* 可评价的，可估价的，按比例的
10. velocity ［viˈlɔsiti］ *n.* 速度；速率；迅速；周转率
11. incline ［inˈklain］ *v.* 使倾斜；赞同；喜爱
12. pitch ［pitʃ］ *n.* （齿轮）节距
13. split ［split］ *v.* 劈开，（使）裂开；分裂，分离　*n.* 裂开，裂口，裂痕
14. pinion ［ˈpinjən］ *n.* 小齿轮
15. overlap ［ˌəuvəˈlæp］ *v.* （与某物）交叠，重叠，重合
16. herringbone ［ˈheriŋbəun］ *n.* 交叉缝式，人字形　*adj.* 人字行的　*v.* （使）成箭尾形
17. vice versa　反之亦然

Reading Material B

Basic Machining Techniques

The importance of machining processes can be emphasized by the fact that every product we use in our daily life has undergone this process either directly or indirectly.

(a) In USA, more than \$100 billion are spent annually on machining and related operations.

(b) A large majority (above 80%) of all the machine tools used in the manufacturing industry have undergone metal cutting.

These facts show the importance of metal cutting in general manufacturing. It is therefore important to understand the metal cutting process in order to make the best use of it.

The five basic techniques of machining metal include drilling and boring, turning, planing, milling, and grinding. Variations of the five basic techniques are employed to meet special situations.

Drilling consists of cutting a round hole by means of a rotating drill. Boring, on the other hand, involves the finishing of a hole already drilled or cored by means of a rotating, offset, single-point tool. On some boring machines, the tool is stationary and the work revolves; on others, the reverse is true.

The lathe, as the turning machine, is commonly called the father of all machine tools. The piece of metal to be machined is rotated and the cutting tool is advanced against it.

Planing metal with a machine tool is a process similar to planing wood with a hand plane. The essential difference lies in the fact that the cutting tool remains in a fixed position while the work is moved back and forth beneath it. Planers are usually large pieces of equipment; sometimes large enough to handle the machining of surfaces 15 to 20 feet wide and twice as long. A shaper differs from a planer in that the workpiece is held stationary and the cutting tool travels back and forth.

After lathes, milling machines are the most widely used for manufacturing applications. Milling consists of machining a piece of metal by bringing it into contact with a rotating cutting tool which has multiple cutting-edges. There are many types of milling machines designed for various kinds of work. Some of the shapes produced by milling machines are extremely simple, like the slots and flat surfaces produced by circular saws. Other shapes are more complex and may consist of a variety of combinations of flat and curved surfaces depending on the shape given to the cutting-edges of the tool and on the travel path of the tool.

Grinding consists of shaping a piece of work by bringing it into contact with a rotating abrasive wheel. The process is often used for the final finishing to close dimensions of a part that has been heat-treated to make it very hard. This is because grinding can correct distortions that may have resulted from heat treatment. In recent years, grinding has also found increased application in heavy-duty metal removal operations.

New Words and Phrases

1. single-point tool   单刃刀具
2. hand plane   手刨, 木工刨
3. planer ['pleinə] *n.* 龙门刨床
4. shaper ['ʃeipə] *n.* 牛头刨床
5. abrasive wheel   砂轮

Reading Material C

Machine Elements

However simple, any machine is a combination of individual components generally referred to as machine elements or parts. Thus, if a machine is completely dismantled, a collection of simple parts remains such as nuts, bolts, springs, gears, cams, and shafts which are the building block of all machinery. A machine element is, therefore a single unit designed to perform a specific function and capable of combining with other elements. Sometimes certain elements are associated in pairs, such as nuts and bolts or keys and shafts. In other instance, a group of elements is combined to form a subassembly, such as bearings, couplings, and clutches.

The most common example of machine elements is a gear, which, fundamentally, is a combination of the wheel and the lever to form a toothed wheel. The rotation of this gear on a hub or shaft drives other gears that may rotate faster or slower, depending upon the number of teeth on the basic wheels.

Other fundamental machine elements have evolved from wheel and lever. A wheel must have a shaft on which it may rotate. The wheel is fastened to the shafts with couplings. The shaft must rest in bearings, may be turned by a pulley with a belt or a chain connecting it to a pulley on a second shaft. The supporting structure may be assembled with bolts or rivets or by welding. Proper application of these machine elements depends upon knowledge of the force on the structure and the strength of the materials employed.

The individual reliability of machine elements becomes the basis for estimating the overall life expectancy of a complete machine.

Many machine elements are thoroughly standardized. Testing and practical experience have established the most suitable dimensions for common structural and mechanical parts. Through standardization, uniformity of practice and resulting economics are obtained. Not all machine parts in use are standardized, however. In the automotive industry, only fasteners, bearings, bushings, chains, and belts are standardized. Crankshafts and connecting rods are not standardized.

New Words and Phrases

1. dismantle [disˈmæntl] vt. 分解（机器），拆开，拆卸
2. nut [nʌt] n. 螺母
3. cam [kæm] n. 凸轮，偏心轮；样板，靠模，仿形板
4. clutch [klʌtʃ] n. 离合器，联轴器；夹紧装置
5. rivet [ˈrivit] n. 铆钉 v. 铆接，铆
6. bushing [ˈbuʃiŋ] n. 衬套；轴衬；轴瓦；［电］（绝缘）套管
7. crank [kræŋk] n. 曲轴

Reading Material D

Milling Machines

Most basic milling machines are of column-and-knee construction, employing the components and motions. The column, mounted on the base, is the main supporting frame for all the other parts and contains the spindle with its driving mechanism. As indicated, this construction provides controlled motion of the worktable in three mutually perpendicular directions: (1) through the knee moving vertically on ways on the front of the column, (2) through the saddle moving transversely on ways on the knee, and (3) through the table, moving longitudinally on ways on the saddle. All of these motions can be imparted either by manual or powered means. In most cases, a powered rapid traverse is provided in addition to the regular feed rates for use in setting up work and in returning the table at the end of a cut.

When a milling machine has provision for only the three mutually perpendicular table motions just described, it is called a plain column-and-knee type. These are available with both horizontal and vertical spindles. A horizontal spindle, plain, column-and-knee type is shown in Fig. II D-1. On this type, an over-arm is mounted on the top of the column to provide an outboard bearing support for the end of the cutter arbor when required. A plain, vertical spindle, column-and-knee milling machine is shown in Fig. II D-2. In some machines of this type, the spindle is mounted in a sliding head that can be fed up and down either by power or by hand. Vertical-spindle machines are especially well suited for face-and-end-milling operations. They also are well suited for drilling and boring, particularly where holes must be accurately spaced in a horizontal plane, because the controlled table motion provides an easy means for accomplishing this.

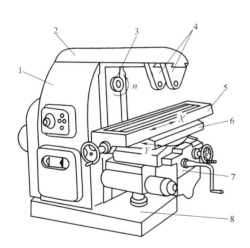

Fig. II D-1 Plain, horizontal spindle,
column-and-knee type milling machine
1—column 2—overarm 3—spindle 4—arbor support
5-table 6—saddle 7—knee 8—base

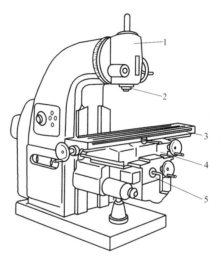

Fig. II D-2 Plain, vertical spindle,
column-and-knee type milling machine
1—head stock 2—spindle
3—table 4—saddle 5—knee

On universal column-and-knee milling machines, a swivel table housing is mounted on the saddle and carries the table. By this feature, the table can be swiveled in a horizontal plane, thus increasing the flexibility of the machine and permitting the milling of helices, as found in twist drills, milling cutters, and helical gear teeth.

Ram-type mulling machine is a special type of column-and-knee milling machine that has a spindle head mounted on the end of a horizontally adjustable ram. This spindle head can be swiveled about a horizontal axis so that milling can be done horizontally, vertically, or at any angle. This added flexibility is advantageous where a variety of work has to be done as in tool and die or experimental shops.

New Words and Phrases

1. column-and-knee 升降台式
2. ram-type 滑枕式
3. outboard ['autbɔːd] *adj.* 外置的
4. helices ['helisiːz] *n.* （pl）螺杆，螺旋状（之物）

Part III Basics of Mechatronics

Unit 9 Electrician Technology

As early as the latter of 16th century, the electrician technology was explored. In the electrical engineering, the most commonly used electrical components include the resistance, the inductance, and the capacitance. The relationship between the electrical components is described by the circuit diagram or network. The analysis of the circuit diagram predicts the performance of the actual device.

The pure resistor element only consumes the electric energy, but it can not reposit the electric energy. The property is defined by the relationship $R = u(t)/i(t)$. If $u(t)$ and $i(t)$ are in volts and amperes respectively, and R is the resistance in ohms, the above expression is known as Ohm's Law.

The circuit element that stores energy in an electric field is a capacitor (also called capacitance). When the voltage is variable over a cycle, energy will be stored during one part of the cycle and returned in the next. While an inductance cannot retain energy after the removal of the source because the magnetic field collapses, the capacitor retains the charge and the electric field can remain after the source is removed. This charged condition can remain until a discharge path is provided, at the same time the energy is released. The charge, $q = Cu$, on a capacitor results in an electrical field in the dielectric which is the mechanism of the energy storage. In the simple parallel-plate capacitor, there is an excess of charges on one plate and a deficiency on the other. It is the equalization of these charges that takes place when the capacitor is discharged. The symbol of the capacitance is usually expressed by C. When u and q are in volts and coulombs respectively, with t in seconds, the units of the capacitance are farads.

The circuit element that stores energy in a magnetic field is an inductor or an inductance. Under the influence of time-variable current, the energy is generally stored during some times of the cycle and then returned to the source during the others. When the inductance is removed from the source, the magnetic field will collapse; in other words, no energy is stored without a connected source. The coils found in electric motors, transformers, and similar devices can be expected to have inductances in their circuit models. The inductance exhibited in a set of parallel conductors must be considered at most frequencies.

In the electrical engineering, a circuit refers to the closed path in which the current is flowing. Because of some certain requirements, the circuit is composed of some electronic devices and components with a certain mode. It is mainly composed of the power supply, load, connecting wire, control and protection devices, and so on.

In the circuit diagram, the power supply can provide energy for the equipment. It can convert the energy from the other forms into the electrical one; the battery and generator all belong to the common powers.

The load refers to a variety of the electrical equipment, which can convert electrical energy into other forms of energy, such as light, electric fan, electric motor, and so on.

The connecting wire can connect the power supply and the load into a closed loop for the power transmission and distribution; the common conductors are copper wires and aluminum wires.

The control and protection devices are used to control the on-off of a circuit in order to protect the circuit, so that the circuit can work normally, such as the switches and fuses.

In the circuit, the charge's directional motion is called the current. In metal conductors, the electric current is the movement of the electronics with a rule in electric field. In some liquids or gases, the current is the movement of the positive, negative ions towards the opposite direction under the electric field force. Current is a physical phenomenon, which equals to the ratio with the amount of charge to the time through the conductor. If the charge amount through conductor cross section is 1 coulomb within one second, the current is the 1 ampere.

Because of the difference between the high potential and the low potential in the currents, the current can flow in the wire. In other words, the potential difference between any two points is called the voltage between the two points in the circuit. The letter U is commonly used for representing the voltage. The voltage promotes the charge's directional movement for the formation of the current.

Today, a series of new science such as electronic and computer technology develop rapidly; all kinds of basic science, applied science and technology develop more closely in the knowledge structure; the different discipline and profession are permeated each other. Therefore, the electrical engineering will be promoted greatly, which can be better for the human life to make greater contribution.

New Words and Phrases

1. diagram ['daiəgræm] n. 图解，简图，图表
2. resistance [ri'zistəns] n. 抵抗，反抗，抵抗能力，电阻，热阻
3. predict [pri'dikt] vt. & v. 预言；预测；预示
4. reposit [ri'pɔzit] vt. 贮藏，使复位
5. removal [ri'mu:vəl] n. 移走，脱掉
6. excess ['ekses] adj. 超重的，过量的；额外的 n. 超过，过多之量
7. respectively [ris'pektivli] adv. 各自地，各个地，分别地
8. distribution [distri'bju:ʃən] n. 分发，分配
9. phenomenon [fi'nɔminən] n. 现象
10. potential [pə'tenʃəl] adj. 潜在的，有可能的 n. 潜力，潜势，可能性
11. contribution [kɔntri'bju:ʃən] n. 捐助物，贡献
12. coil [kɔil] n. （一）卷，（一）圈；盘卷之物；线圈；簧圈；盘管

Notes

1\. In the electrical engineering, the most commonly used electrical components includes the resistance, the inductance, and the capacitance. The relationship between the electrical components is described by the circuit diagram or network. The analysis of the circuit diagram predicts the performance of the actual device.

参考译文：在电工学中，常用的电气元件有电阻元件、电感元件和电容元件，它们之间的关系是通过电路图或网络来描述的，对电路图的分析可预估实际器件的性能。

2\. Under the influence of time-variable current, the energy is generally stored during some times of the cycle and then returned to the source during others.

参考译文：在时变电流的作用下，电感在一个周期的一段时间里存储能量，而在其他时间段里又释放能量给电源。

3\. Current is a physical phenomenon, which equals to the ratio with the amount of charge to the time through the conductor. If the charge amount through conductor cross section is 1 coulomb within one second, the current is the 1 ampere.

参考译文：电流是一种物理现象，在数值上等于通过导体横截面的电荷量 q 和通过这些电荷量所用时间的比值。如果在1s内通过导体横截面的电荷量是 $1C$，导体中的电流为 $1A$。

Exercises

1\. Answer the following questions according to the text.
a) What's the inductance and its function?
b) What is the capacitance?
c) In the circuit diagram, what is the function of the control and protection devices?

2\. Put the following words into Chinese by reference to the text.
describe   relationship   expression   retain
conductor   protection   formation   equal

3\. Translate the following sentences into English.
a) 电器是指能根据外界特定的信号和要求，自动或手动地接通与断开电路，断续或连续地改变电路参数，实现对电路的切换、控制、保护、检测和调节用的电气设备。
b) 普通导电材料是指专门用来传导电流的金属材料。

Unit 10   Electron Technology

Electron technology is a kind of rising technology in the end of the nineteenth century. It has the most rapid development and been the most widely used in modern science and technology in the beginning of the twentieth century. It has become an important symbol of new technology. Based on the principle of electronics, electronic technology applies electronic devices to design and manufacture some specific function circuit for solving the practical problems of science. Electronic technology is mainly used in electronic signal processing. The main processing methods are signal generation, amplification, filtering and conversion.

The first generation of electronic product is based on electronic tube. In the late of the 1840s, the first semiconductor transistor was borned. It is compact, lightweight, energy-saving, long life, and soon replace the tube. In the end of the 1950s, the first integrated circuit appeared. It puts a lot of transistors and other electronic components integrated on a silicon chip, and make electronic products smaller. IC has the rapid development for the LSI and the VLSI, so that the directions of development of the electronic products are high-performance, low consumption, high accuracy, high stability, intelligent direction, and so on. In general, electronic technology is divided into analog electronic technology and digital electronics technology.

The conventional analog electronic circuit is mainly composed of some electronic devices, such as rectifier circuit, amplifier circuit, oscillator circuit, transform circuit, and other basic electronic devices, as well as some basic function circuits which is composed of the various uses of the device or system. The conventional analog electronic circuit is the main circuit in the form of an electronic device. The electronic circuit has been widely used in telecommunications, broadcasting, television, computer, and industrial control. In the transmission of information, it has the performances of rapid, sensitive, accurate, easy to implement remote control, compact, reliable, and easy to use, so the electronic circuit becomes the extremely important part of the modern and advanced science and technology.

The main features of the analog amplification circuit are as follows.

- The processed signals are continuously varying analog signals, such as audio signals, television pulse signal, temperature, pressure changes, and so on.
- The triode in the analog circuit acts as a zoom component, rather than in the digital circuit, the triode acts as a switch.
- Analysis methods mainly adopt the graphic method and the small signal equivalent circuit method to analyze the static and dynamic working conditions of the amplification circuits. In the digital circuits, the logic algebra, the truth table, Karnaugh map, and state transformation graph are often used to analyze the logical relationships between the input and output.

Generally, the digital signal is referred to the discrete electrical signal changing in time and the values. This change are occurring in a series of discrete moments, that is, the signal always changes back and forth between high level and low level. The digital circuit is used to process the digital signal of the electronic circuit. The processing operations include the digital signal transmission, logic operation, control, counting, storage, display, pulse waveform generation and transformation, and so on.

Modern digital circuits are constructed from a number of digital integrated device made by the semiconductor process. Modern digital circuits only transmit two state information (namely 0 and 1) for the logical operations. Its main features are as follows.

- The semiconductor devices of the digital circuits are working in the switch state.
- The signals of the digital circuits are discontinuous changing pulse signals.
- The analysis methods of the digital circuits commonly include the logic algebra, the truth ta-

bles, Karnaugh map, the status transformation map, and so on.

Because the digital circuits have the advantages of simple structure, easy manufacture, low cost, reliable work, and so on, they have been widely used in automatic control, measuring instruments, communications, science and technology.

In recent years, the rapid progress of the programmable logic device (PLD) and especially the field programmable gate array (FPGA), give the digital electronic technology a new situation, larger scale, the combination of hardware and software, more perfect function, and more flexible use.

New Words and Phrases

1. symbol ['simbəl] n. 象征，标志
2. amplification [æmpləfi'keiʃən] n. 扩大
3. filter ['filtə] n. 过滤，过滤器 vt. & v. 透过，过滤
4. conversion [kən'və:ʃən] n. 变换，转化
5. lightweight ['laitweit] adj. 轻量的，薄型的 n. 轻量
6. consumption [kən'sʌmpʃən] n. 消费，消耗；消费［耗］量
7. analog ['ænəlɔ:g] n. 类似物；同源语；模拟 adj. （钟表）有长短针的；模拟的
8. implement ['implimənt] vt. 使生效，贯彻，执行 n. 工具，器具，用具
9. extremely [iks'tri:mli] adv. 极端；极其；非常
10. algebra ['ældʒibrə] n. 代数学，代数
11. reliable [ri'laiəbl] adj. 可靠的，可信赖的
12. logic ['lɔdʒik] n. 逻辑（学），逻辑性

Notes

1. Based on the principle of electronics, electronic technology applies electronic devices to design and manufacture some specific function circuit for solving the practical problems of science.

参考译文：电子技术根据电子学的原理，运用电子器件设计和制造某种特定功能的电路从而解决实际问题。

2. The electronic circuit has been widely used in telecommunications, broadcasting, television, computer, and industrial control. In the transmission of information, it has the performances of rapid, sensitive, accurate, easy to implement remote control, compact, reliable, and easy to use, so the electronic circuit becomes the extremely important part of the modern and advanced science and technology.

参考译文：电子电路在通信、广播、电视、计算机和工业控制等方面得到广泛的应用。它具有传递信息快速、灵敏、精确、容易实现遥控，而且体积小巧、运行可靠、使用方便的优点，是实现现代化先进科学技术的极其重要的一个组成部分。

3. The triode in the analog circuit acts as a zoom component, rather than in the digital circuit, the triode acts as a switch.

参考译文：晶体管在电路中的作用相当于一个放大器件，而不像在数字电路中晶体管的作用相当于一个开关。

Exercises

1. Answer the following questions according to the text.
a) What's the developing direction of the electronic technology?
b) Which links are composed of analog electronic circuit?
c) What's the main characteristic of the digital circuit?
2. Put the following words into Chinese by reference to the text.
transformation   structure   manufacture   communication
programmable   combination   dynamic   remote
3. Translate the following sentences into English.

a) 各种电子电路及系统均需要直流电源供电，大多数是利用电网的交流电源经过变换而获得的。

b) 放大电路的功能是把微弱的电信号不失真地放大到所需的数值。

Unit 11 Automatic Control Systems

A large impetus to the theory and practice of automation control occurred during World War II. The automatic control system is the production process or other process carried out according to the desired rules or predetermined program in no one direct participation. The automatic control system is the main way of realizing automation. The automatic control system is composed of the controller, the controlled object, the implementing agencies, and the transmitter. Automatic control systems have been widely used in different fields of human society.

In industry, such as metallurgy, chemical industry, machinery manufacturing, and other production processes, the various physical quantities including temperature, flow, pressure, thickness, tension, speed, location, frequency, phase, etc., have the corresponding control system. On these basis, not only the digital control systems with the better control performance and higher degree of automation, but also the process control system with the double functions of control and management are established. In agriculture, the water level automatic control system and the automatic operation system of agricultural machinery belong to the scope of application of the automatic control system. Process control is concerned with maintaining at a desired value of processing variables such as temperature, pressure, flow rate, liquid level, viscosity, density, and composition.

Much current work in process control involves extending the use of the digital computer to provide direct digital control (DDC) of the manipulated variables. Computers are now used to control many types of devices such as manufacturing processes, security systems, burglar alarms, air conditioning and central heating systems in large building.

In military technology, automatic control is usually applied to various types of servo systems, fire control systems, guidance and control systems. In the aerospace, aviation, and navigation, in addition, the application area also includes a navigation system, remote control system, and all kinds of simulator.

In addition, in office automation, library management, traffic management, and daily life, the automatic control technology also has the practical application. With the development of control theory and control technology, automatic control system applications are still expanding, almost involved in biological, medical, ecological, economic, social, and all other fields. Next, we discuss a few examples of modern industry.

The servomechanism is commonplace in automatic control. The servomechanism, or' servo' for short, is a closed-loop control system in which the controlled variable is mechanical position or motion. It is designed so that the output will quickly and precisely respond to a change in the input command. Therefore, we also think of a servomechanism as a following device. Another form of servomechanism in which the rate of change or velocity of the output is controlled is known as a rate or velocity servomechanism.

The energy conversion and distribution are very important in electric power. Large modern power plants which may exceed several hundred megawatts of generation require complex control systems to account for the interrelationship of the many variables and provide optimum power production. Control of power generation may be generally regarded as an application of process control, and it is common to have as many as 100 manipulated variables under the computer control.

Automatic control has also been extensively applied to the distribution of electric power. Power systems are commonly made up of a number of generating plants. As load requirements fluctuate, the generation and transmission of power is controlled to achieve minimum cost of system operation. In addition, most large power systems are interconnected with each other, and the flow of power between systems is controlled.

In the machining process, there are a lot of manufacturing operations such as boring, drilling, milling, and welding which must be performed with high precision on a repetitive basis. Numerical control is a system that uses predetermined instructions called a program to control a sequence of such operation. The instructions to accomplish a desired operation are coded and stored on some medium such as punched paper tape, magnetic tape, or punched cards. These instructions are usually stored in the form of numbers, hence the name is numerical control. The instructions identify which tool is to be used, in which way, and the path of the tool movement.

To provide mass transportation systems for modern urban areas, large, complex control systems are needed. Several automatic transportation systems now in operation have high-speed trains running at several-minute intervals. Automatic control is necessary to maintain a constant flow of trains and to provide comfortable acceleration and braking at station stops.

Aircraft flight control is another important application in the transportation field. This has been proven to be one of the most complex control applications due to the wide range of system parameters and the interaction between controls. Aircraft control systems are frequently adaptive in nature; that is, the operation adapts itself to the surrounding conditions. For example, since the behavior of an aircraft may differ radically at low and high altitudes, the control system must be modified as a function of altitude. Ship-steering and roll-stabilization controls are similar to flight control, but generally require far higher powers and involve lower speeds of response.

New Words and Phrases

1. impetus ['impitəs] *n.* 推动，促进，刺激
2. predetermine [pri:di'tɜ:min] *vt. & v.* 预先裁定
3. participation [pɑ:tisə'peiʃən] *n.* 参加，参与
4. viscosity [vi'skɔsiti:] *n.* <术> 黏稠；黏性
5. density ['densiti] *n.* 密集，稠密；〈物〉〈化〉密度
6. composition [kɔmpə'ziʃən] *n.* 构图；构成，成分
7. variable ['vɛəriəbl] *adj.* 变化的，可变的，易变的 *n.* 可变因素；变数
8. burglar ['bɜ:glə] *n.* 窃贼，破门盗窃者
9. fluctuate ['flʌktjueit] *v.* 波动，涨落，起伏
10. behavior [bi'heivjə] *n.* 行为，举止；态度

Notes

1. The automatic control system is the production process or other process carried out according to the desired rules or predetermined program in no one direct participation.

参考译文：自动控制系统是在无人直接参与下可使生产过程或其他过程按期望规律或预定程序进行的控制系统。

2. In industry, such as metallurgy, chemical industry, machinery manufacturing, and other production processes, the various physical quantities including temperature, flow, pressure, thickness, tension, speed, location, frequency, phase, etc., have the corresponding control system.

参考译文：冶金、化工、机械制造等工业生产过程中的各种物理量，包括温度、流量、压力、厚度、张力、速度、位置、频率、相位等，都有相应的控制系统。

3. Ship-steering and roll-stabilization controls are similar to flight control, but generally require far higher powers and involve lower speeds of response.

参考译文：船舶转向和颠簸稳定控制与飞行控制相似，但是一般需要更大的功率和较低的响应速度。

Exercises

1. Answer the following questions according to the text.

a) What are the parts of the automatic control system?

b) What's the effect of the digital computer's introduction to the automatic control system?

c) Please give some example of the automatic control system's application.

2. Put the following words into Chinese by reference to the text.

carry out   desire   be composed of   belong to

apply to   servomechanism   for short   acceleration

3. Translate the following sentences into English.

a) 飞机自动驾驶仪是一种能保持或改变飞机飞行状态的自动装置。它可以稳定飞行的姿态、高度和航迹。

b) 自动控制系统利用外加的设备或装置使机器、设备或生产过程的某个工作状态或参数自动地按照预定的规律进行修正。

Unit 12 Mechatronics Systems

With the development of producing technology and the penetration of microelectronic technology, automation technology to the field of mechanical technology, a new field, mechatronics technology is formed. On the one hand, mechatronics technology greatly improves the product's performance and the market's competitiveness; on the other hand, it greatly enhances the adaptability of the products to the environment and human activities expanding to the space. For example, the United States of America's Apollo landing on the moon and China's Shenzhou 5, E 1, E 2 are the result of the development of mechatronics technology. Because of the tremendous impetus to the development of modern industry and technology, mechatronics technology causes each country's great attention in the world.

The English term of "mechatronics" is originated in Japan, which makes the first half of the mechanics (mechanics) and the second half of the electronics (electronics) into a new word, and it means the synthesis of the two disciplines of mechanics and electronics. However, the "mechatronics" is not a simple overlay of the mechanical technology and electronic technology, it blends electronic technology, information technology, and automatic control functions into the mechanical device. With the organic integration of various technologies, the product's performance will be the best.

With the development of science and technology, the organic combination of mechanical part and the electronic part makes mechanical and electrical products to achieve their optimal performance on the point of view of system. The basic concept of mechatronics can be summarized as follows: from the viewpoint of system, the organic synthesis of mechanical technology, microelectronics, information technology, control technology, computer technology, sensor technology, and interface technology in system engineering to realize the whole system optimization, so as to be a new science and technology.

The so-called mechatronics contains two aspects of mechatronics technology and mechatronics systems. Mechatronics technology includes technical basis and technical principle, and it makes mechatronics system to realize, use and develop. Mechatronics system also includes the mechatronics products and mechatronics production system. Because of applying the mechatronics technology into mechatronics equipment according to the objectives and requirements, mechatronics production systems have the advantages of high productivity, high-quality, high reliability, high flexibility, and low-power. For example, the flexible manufacturing systems (FMS), computer-aided design and manufacturing (CAD/CAM), computer-aided process planning (CAPP), computer integrated manufacturing system (CIMS), and a variety of industrial process control systems all belong to the mechatronics production systems. The new generation of products or equipment with the mechatronics features or mechatronics technologies are collectively referred to as the mechatronics products.

Mechatronics products and systems have been infiltrated into the national economy and the daily life and work. Refrigerators, automatic washing machines, VCRs, cameras and other household appliances, electronic typewriters, copiers, fax machines and other office automation equipment, industrial robotics, automated material handling vehicles, magnetic resonance imaging diagnostic equipment and other machinery equipment all belong to the mechatronics products.

In order to continuously meet the diverse requirements of people's lives and productions needs of labor-saving, time-saving, automation, mechatronics products are constantly innovated. With the advantage of modern high technology, mechatronics technology have made a more significant benefits in technology, economic, society and other fields such as improving the accuracy, enhancing the functions, improving the operability and practicality, improving productivity, reducing costs, saving energy, reducing consumption, reducing labor intensity, improving the safety and reliability, improving labor conditions, simplifying the structure, reducing the labor intensity, improving working conditions, improving the safety and the reliability, simplifying structure, reducing weight to enhance the flexibility and intelligent, reducing the price, and so on. And it is promoting the progress of society, science and technology. Overall, with the features of multi-function, high efficiency, high intelligence, and high reliability, with the advantages of lightness, thinness, and compactness in appearance, mechatronics products have been widely used in all aspects of the production and daily-life.

Mechatronics system is generally composed of the mechanical part, the sensor detection section, the dynamic and driving execution component, the control system, and information processing section. The five components achieve motion transmission, information control, and energy conversion within them and between them through the interface coupling. Among them, the mechanical parts of the system will support other components; the power systems will provide power for the system's normal operation; the sensor detection part will test a variety of parameters and status producing in the system and the external environment when the system is operating, and it can convert the signals to be identified and transmit the signals to the information processing unit; the control system will centralize, storage, analysis, and process the sensor's detecting informations and the external input commands, according to the results of information processing, and the control system will issue the corresponding commands so as to control the whole system orderly work in accordance with certain procedures and rhythm; the driving components will execute the corresponding actions according to the control information and instruction. The constituent elements of mechatronics system make it have five major functions of control, detection, power, action, and structure. The five major parts have the different roles on their different duties.

New Words and Phrases

1. tremendous [tri'mendəs] *adj.* 极大的，巨大的
2. attention [ə'tenʃən] *n.* 注意，专心，留心
3. originate [ə'ridʒineit] *v.* 起源于，来自，产生 *vt.* 创造，创始，开创；发明
4. synthesis ['sinθisis] *n.* 综合，综合法；〈化〉合成

5. discipline ['disiplin] vt. 训练，训导 n. 训练，学科
6. blend [blend] vt. & v. （使）混合，（使）混杂 n. 混合物
7. optimal ['ɔptəməl] adj. 最佳的，最优的
8. performance [pə'fɔːməns] n. 演出，表演，性能，工作情况
9. viewpoint ['vjuːpɔint] n. 观点，意见，角度
10. optimization [ɔptimai'zeiʃən] n. 最佳化，最优化
11. flexibility [fleksi'biliti] n. 柔韧性；机动性，灵活性
12. generation [dʒenə'reiʃən] n. 同时代的人，一代人，一代
13. resonance ['rezənəns] n. 回响，回荡；洪亮；共鸣
14. transmission [trænz'miʃən] n. 传送，传播，传达；播送

Notes

1. With the development of science and technology, the organic combination of mechanical part and the electronic part makes mechanical and electrical products to achieve its optimal performance on the point of view of system.

参考译文：随着科学技术的发展，机电一体化产品把机械部分与电子部分有机结合，从系统的观点使其达到最优化。

2. In order to continuously meet the diverse requirements of people's lives and production effort, time-saving, automation needs, mechatronics products are constantly innovated.

参考译文：为了不断满足人们生活的多样化要求和生产的省力、省时和自动化等方面的需要，机电一体化产品不断推陈出新。

3. Mechatronics system is generally composed of the mechanical part, the sensor detection section, the dynamic and driving execution component, the control system, and information processing section. The five components achieve motion transmission, information control, and energy conversion within them and between them through the interface coupling.

参考译文：机电一体化系统一般由机械部分、传感检测部分、动力及驱动执行部件、控制系统及信息处理等部分组成，这些组成要素内部及其之间，通过接口耦合来实现运动传递、信息控制、能量转换。

Exercises

1. Answer the following questions according to the text.
a) Please give the description (the producing process) of mechatronics.
b) What's the developing direction of the mechatronics?
2. Put the following words into Chinese by reference to the text.
mechatronics   overlay   integration   principle
high-quality   feature   collectively   diagnostic
3. Translate the following sentences into English.
a) 随着科学技术的发展，机电一体化技术及机电一体化产品日新月异，遍布生活的每一个角落。

b）机电一体化是一个综合的概念，机电一体化产品具有较高的技术含量，其技术附加值随机电结合程度的加深而提高。

Reading Material A

Alternating Current

In a very short time, the alternating current intensity starts at zero, increases to maximum strength, and then falls to zero. The current then flows in the opposite direction, its strength will increase from zero to maximum and then decrease to zero. In this manner, the current changes back and forth. A current that changes its direction of flow at regular intervals is called alternating current (AC). The graphical representation of the variation of an alternating current plotted as a function of time is called the waveform of the current. Generally, the period is represented by T and measured in second. The reciprocal of the value T is known as the frequency and is defined as the number of periods occurring in the unit of time. We usually express frequency in hertz or cycles per second.

An alternating quantity is called symmetrical when all values separated by a half period have the same magnitude but opposite sign. The half period average value of a symmetrical alternating current is the average of the value of current taken throughout a half period beginning with a zero value. The average value of an alternating current is of limited used and it is principally employed only when the AC wave must be rectified. The root mean square (r. m. s.) or effective value of an alternation current is more useful. This value is equal to the square root of the mean value of the squares of the instantaneous values taken throughout one cycle. It is not difficult to demonstrate that the effective value of an alternating current is the number of amperes of direct current that would heat a given resistor at the same average rate as that at which the alternating current heats it.

In practice, the waves of the current and the voltage are essentially different from the sine wave, but in the majority of cases these waves are sufficiently near to a sine wave. Therefore, when we consider the waveform as the sine wave, the results we obtain and the sinusoidal waves are precise enough for practical purposes.

New Words and Phrases

1. desirable [diˈzaiərəbl] *adj.* 想要的，值得要的
2. consideration [kənˌsidəˈreiʃən] *n.* 考虑，思考
3. situation [sitjuˈeiʃən] *n.* 位置，地点
4. interval [ˈintəvəl] *n.* 间隔，时间（间隔）
5. at regular intervals 每隔一段时间
6. fraction [ˈfrækʃən] *n.* 小部分，片段，分数
7. maximum [ˈmæksiməm] *n.* 最大值，最大量，最大限度

Reading Material B

Digital Television

Even since Philo T. Farnsworth put together the first television set in Indiana, in 1927, there have been only two major changes in the way TV sets working processes: one is, 1954 references color; secondly, 1970s from the tube to the transistor. Now a profound change is about to occur, which is digital television. With the different method for signal transmission, digital television will greatly change the future TV screen and working mode.

The digital television is a cross between a computer terminal and a TV set. Although the changes it will bring may not be dramatic, its improving quality will be increasingly appreciated. The effects such as the zoom effects, stereo sound, and freeze-frames views of live shoes will become commonplace. Digital TV has hopes to give viewers a clearer, more consistent picture than that has been available so far.

Since the 1950s, almost all the volume of electronic products have been gradually narrowing. Calculators now slide into checkbooks, stereo speakers no larger than bricks pack an audio wallop, and compact discs have compressed the record industry. In the course of reform, people's television is still very large, this is mainly because of the volume of one of the television's primary components——CRT more difficult and uneconomical to reduce.

MRS Technology Inc. has developed a lithographic production system specifically for the high volume manufacture of flat panel, active matrix liquid crystal displays. The model, 4500 panel printer, as the machine is called can make one inchthick color LCD screens up to 18 inches diagonal in size. The weigh is only a few pounds and it can be hung on a wall like a painting.

New Words and Phrases

1. Indiana ['indi'ænə] *n.* 印第安纳
2. digital ['didʒitəl] *adj.* 数字的
3. sale [seil] *n.* 卖,出售
4. on sale 出售的
5. cross [krɔs] *n.* 混合种
6. zoom [zu:m] *v. n.* 图像放大;变焦距
7. dramatic [drə'mætik] *adj.* 戏剧的,惊人的,奇迹般的

Reading Material C

Adaptive Control Systems

An adaptive control system is the one whose parameters are automatically adjusted to compensate for corresponding variations in the properties of the process. Naturally, there must be some criteria on which an adaptive program is based. To set a value for the controlled variable is not enough.

To meet this target, not only the adaptive control is needed, but also the "objective function" must be additionally provided.

The objective function for a given process may be the damping of the controlled variable. Therefore, there are two loops, one operating on the controlled variable, the other operating on its damping. Because of the damping marking the dynamic loop gain, the system is called the dynamic adaptive system. It is also possible to stipulate an objective function of the steady-state gain of the process. This control system is designed to the steady-state adaptive system.

The prime function of the dynamic adaptive systems is to give a control loop a consistent degree of stability. So the dynamic loop gain is then the objective function of the controlled variable being regulated; its value is to be specified. The value of the manipulated variable which can satisfy the objective function is known relative to the major cases prevailing within the process, therefore, it can be easily compiled program for adaptive control. For example, the optimum fuel-air ratio may be known for various conditions of air flow and temperature. With the changes of the controller setting values to design the control system, the ratio of fuel-air can adapt to the change of the ratio of the air flow-temperature. So it can be a function of flow in the dynamic adaptation.

The dynamic adaptive system controls the dynamic gain of a loop, so its counterpart seeks a constant steady-state process gain. Of course, this means that the static process gain is the changing, and one specific value is desired.

New Words and Phrases

1. adaptation [ædæp'teiʃən] n. 适合，适应，适应性控制
2. designate ['dezigneit] vt. 指明，称为，标志
3. confusion [kən'fjuːʒən] n. 混乱，混淆
4. prevail [pri'veil] v. 流行，经常发生

Reading Material D

Prospect of New Mechatronics Products

With the development of science and technology, more and more new mechatronics products appear. For example, the robot belongs to the intelligent mechatronics products. When you see the robots doing homework on television, you may not be aware of the robot already existing in your life; But, according to the definition of robot, washing machine, electric heaters, etc., all belong to the robots. Robots can also be designed to do the dangerous work in research laboratories or in outer space.

All the satellites launched into outer space have had robots on board with the way of radio, these robots can sent the important information such as temperature and radiation back to their masters on earth. The robot can even taken photographs of the earth and other planets from the high position in space.

The first spaceship has landed on Mars and Venus with the robots rather than human being.

Robots can map the surfaces of the heavenly bodies, make necessary geological studies, and explore unknown places. Over the past two decades, industry has realized that the productivity must be increased and the manufacturing costs must be reduced in order to be competitive in the world markets. Because the skilled workers are gradually decreased, and fewer and fewer people are willing to engage in the heavy, monotonous, poor environment, the industry finds it necessary to automate the many manufacturing processes. The development of computer has made it possible for industry to reliable machine tools and robots, which are making manufacturing processes more productive and reliable, so as to improve their product's competitive position in the world market.

New Words and Phrases

1. robot ['rəubət] n. 机器人
2. gradually ['grædjuəli] adv. 逐步地，渐渐地
3. satellite ['sætəlait] n. 卫星
4. competitive [kəm'petitiv] adj. 竞争的，比赛的
5. geological [dʒiə'lɔdʒikəl] adj. 地质（学）的

Part IV Mechatronics Technique

Unit 13 Introduction to AT89S51

Product Overview

The AT89S51 is a low-power, high-performance CMOS 8-bit microcontroller with 4K bytes of In-System Programmable Flash Memory. The device is manufactured using ATMEL's high-density nonvolatile memory technology and is compatible with the industry-standard 80C51 instruction set and pinout. The on-chip Flash allows the program memory to be reprogrammed in-system or by a conventional nonvolatile memory programmer. By combining a versatile 8-bit CPU with In-System Programmable Flash on a monolithic chip, the AT89S51 is a powerful microcontroller which provides a highly-flexible and cost-effective solution to many embedded control applications.

Features

- Compatible with MCS® -51 Products
- 4K Bytes of In-System Programmable (ISP) Flash Memory
    - Endurance: 1000 Write/Erase Cycles
- 4.0V to 5.5V Operating Range
- Fully Static Operation: 0Hz to 33MHz
- Three-level Program Memory Lock
- 128 ×8-bit Internal RAM
- 32 Programmable I/O Lines
- Two 16-bit Timer/Counters
- Six Interrupt Sources
- Full Duplex UART Serial Channel
- Low-power Idle and Power-down Modes
- Interrupt Recovery from Power-down Mode
- Watchdog Timer
- Dual Data Pointer
- Power-off Flag
- Fast Programming Time
- Flexible ISP Programming (Byte and Page Mode)

Description

The AT89S51 provides the following standard features: 4K bytes of Flash, 128 bytes of RAM, 32 I/O lines, watchdog timer, two data pointers, two 16-bit timer/counters, a five-vector two-level interrupt architecture, a full duplex serial port, on-chip oscillator, and clock circuitry. In addition, the AT89S51 is designed with static logic for operation down to zero frequency and supports two software selectable power saving modes. The Idle Mode stops the CPU while allowing the RAM, timer/counters, serial port, and interrupt system to continue functioning. The Power-down Mode saves the RAM contents but freezes the oscillator, disabling all other chip functions until the next external interrupt or hardware reset.

Pin Configurations

PDIP and PLCC pin configurations are shown in Fig. 13-1a and b.

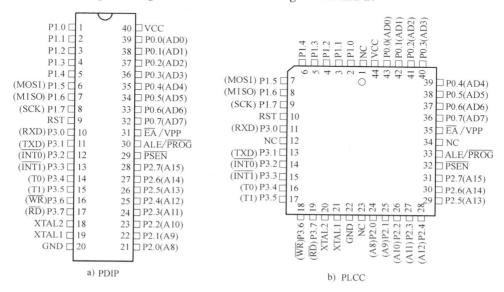

Fig. 13-1   PDIP and PLCC pin configurations

Pin Description

VCC

Supply voltage.

GND

Ground.

Port 0

Port 0 is an 8-bit open drain bi-directional I/O port. As an output port, each pin can sink eight TTL inputs. When 1s are written to Port 0 pins, the pins can be used as high-impedance inputs.

Port 0 can also be configured to be the multiplexed low-order address/data bus during accesses to external program and data memory. In this mode, Port 0 has internal pull-ups.

Port 0 also receives the code bytes during Flash programming and outputs the code bytes during program verification. External pull-ups are required during program verification.

Port 1

Port 1 is an 8-bit bi-directional I/O port with internal pull-ups. The Port 1 output buffers can sink/source four TTL inputs. When 1s are written to Port 1 pins, they are pulled high by the internal pull-ups and can be used as inputs. As inputs, Port 1 pins that are externally being pulled low will source current (I_{IL}) because of the internal pull-ups.

Port 1 also receives the low-order address bytes during Flash programming and verification.

Port Pin	Alternate Functions
P1.5	MOSI (used for In-System Programming)
P1.6	MISO (used for In-System Programming)
P1.7	SCK (used for In-System Programming)

Port 2

Port 2 is an 8-bit bi-directional I/O port with internal pull-ups. The Port 2 output buffers can sink/source four TTL inputs. When 1s are written to Port 2 pins, they are pulled high by the internal pull-ups and can be used as inputs. As inputs, Port 2 pins that are externally being pulled low will source current (I_{IL}) because of the internal pull-ups.

Port 2 emits the high-order address byte during fetches from external program memory and during accesses to external data memory that use 16-bit addresses (MOVX @ DPTR). In this application, Port 2 uses strong internal pull-ups when emitting 1s. During accesses to external data memory that use 8-bit addresses (MOVX @ Ri), Port 2 emits the contents of the P2 Special Function Register.

Port 2 also receives the high-order address bits and some control signals during Flash programming and verification.

Port 3

Port 3 is an 8-bit bi-directional I/O port with internal pull-ups. The Port 3 output buffers can sink/source four TTL inputs. When 1s are written to Port 3 pins, they are pulled high by the internal pull-ups and can be used as inputs. As inputs, Port 3 pins that are externally being pulled low will source current (I_{IL}) because of the internal pull-ups.

Port 3 receives some control signals for Flash programming and verification.

Port 3 also serves the functions of various special features of the AT89S51, as shown in the following table.

Port Pin	Alternate Functions
P3.0	RXD (serial input port)
P3.1	TXD (serial output port)
P3.2	$\overline{INT0}$ (external interrupt 0)
P3.3	$\overline{INT1}$ (external interrupt 1)
P3.4	T0 (timer 0 external input)
P3.5	T1 (timer 1 external input)
P3.6	$\overline{WR}$ (external data memory write strobe)
P3.7	$\overline{RD}$ (external data memory read strobe)

RST

Reset input. A high on this pin for two machine cycles while the oscillator is running resets the device. This pin drives High for 98 oscillator periods after the Watchdog timer times out. The DISRTO bit in SFR AUXR (address 8EH) can be used to disable this feature. In the default state of bit DISRTO, the RESET HIGH out feature is enabled.

ALE/PROG

Address Latch Enable (ALE) is an output pulse for latching the low byte of the address during accesses to external memory. This pin is also the program pulse input ($\overline{PROG}$) during Flash programming.

In normal operation, ALE is emitted at a constant rate of 1/6 the oscillator frequency and may be used for external timing or clocking purposes. Note, however, that one ALE pulse is skipped during each access to external data memory.

If desired, ALE operation can be disabled by setting bit 0 of SFR location 8EH. With the bit set, ALE is active only during a MOVX or MOVC instruction. Otherwise, the pin is weakly pulled high. Setting the ALE-disable bit has no effect if the microcontroller is in external execution mode.

PSEN

Program Store Enable ($\overline{PSEN}$) is the read strobe to external program memory. When the AT89S51 is executing code from external program memory, $\overline{PSEN}$ is activated twice each machine cycle, except that two $\overline{PSEN}$ activations are skipped during each access to external data memory.

EA/VPP

External Access Enable. $\overline{EA}$ must be strapped to GND in order to enable the device to fetch code from external program memory locations starting at 0000H up to FFFFH. Note, however, that if lock bit 1 is programmed, $\overline{EA}$ will be internally latched on reset. $\overline{EA}$ should be strapped to VCC for internal program executions.

This pin also receives the 12-volt programming enable voltage (VPP) during Flash programming.

XTAL1

Input to the inverting oscillator amplifier and input to the internal clock operating circuit.

XTAL2

Output from the inverting oscillator amplifier.

New Words and Phrases

1. CMOS (Complementary Metal Oxide Semiconductor) 互补金属氧化物半导体
2. nonvolatile [nɔn'vɔlətail] *adj.* 非易失性的
3. In-System Programmable Flash Memory 在系统可编程序的 Flash 只读存储器
4. compatible [kəm'pætəbl] *adj.* 适合的；相容的
5. instruction set 指令集
6. pinout ['pinaut] *n.* 引出线
7. programmer ['prəuˌgræmə] *n.* 程序设计者；程序设计器

8. versatile ['və:sətail] *adj.* （指工具、机器等）多用途的；多才多艺的；多功能的
9. monolithic [ˌmɔnə'liθik] *adj.* 独块巨石的；整体的；庞大的
10. microcontroller [ˌmaikrəkən'trəulə] *n.* 微控制器
11. flexible ['fleksəbl] *adj.* 灵活的；易弯曲的；柔韧的；易被说服的
12. embedded [em'bedid] *adj.* 植入的，深入的，内含的
13. watchdog timer 监视计时器 [WDT]
14. oscillator ['ɔsileitə] *n.* 振荡器
15. multiplexed ['mʌltiˌpleksid] *adj.* 多路复用的
16. verification [ˌverəfi'keiʃən] *n.* 证明；证实；核实
17. strobe [strəub] *n.* 闸门，起滤波作用

Notes

1. The device is manufactured using ATMEL's high-density nonvolatile memory technology and is compatible with the industry-standard 80C51 instruction set and pinout.

参考译文：器件采用ATMEL公司的高密度非易失性存储技术生产，兼容工业标准80C51的指令系统及引脚。

2. In addition, the AT89S51 is designed with static logic for operation down to zero frequency and supports two software selectable power saving modes.

参考译文：同时，AT89S51可降至0Hz的静态逻辑操作，并支持两种软件可选的节电工作模式。

3. The Power-down Mode saves the RAM contents but freezes the oscillator, disabling all other chip functions until the next external interrupt or hardware reset.

参考译文：掉电方式保存RAM中的内容，但振荡器停止工作并禁止其他所有部件工作直到下一个外部中断或硬件复位。

4. Port 0 can also be configured to be the multiplexed low-order address/data bus during accesses to external program and data memory.

参考译文：在访问外部数据存储器或程序存储器时，P0口线分时转换成地址总线（低8位）和数据总线复用。

5. Address Latch Enable (ALE) is an output pulse for latching the low byte of the address during accesses to external memory.

参考译文：当访问外部程序存储器或数据存储器时，ALE（地址锁存允许）输出脉冲用于锁存地址的低8位字节。

6. Program Store Enable (PSEN) is the read strobe to external program memory.

参考译文：程序储存允许（PSEN）输出是外部程序存储器的读选通信号。

Exercises

1. Answer the following questions according to the text.

a) The AT89S51 is an 8-bit microcontroller manufactured by Atmel Corporation. Yes or no?

b) The AT89S51 has 4 bi-directional I/O ports, 4K bytes of In-System Programmable program memory. Yes or no?

2. Please introduce the components of AT89S51 microcontroller.

3. Translate the following sentences into Chinese.

a) The AT89S51 implements 128 bytes of on-chip RAM. The 128 bytes are accessible via direct and indirect addressing modes. Stack operations are examples of indirect addressing, so the 128 bytes of data RAM are available as stack space.

b) The WDT is intended as a recovery method in situations where the CPU may be subjected to software upsets. The WDT consists of a 14-bit counter and the Watchdog Timer Reset (WDTRST) SFR. The WDT is defaulted to disable from exiting reset. To enable the WDT, a user must write 01EH and 0E1H in sequence to the WDTRST register (SFR location 0A6H). When the WDT is enabled, it will increment every machine cycle while the oscillator is running. The WDT timeout period is dependent on the external clock frequency. There is no way to disable the WDT except through reset (either hardware reset or WDT overflow reset). When WDT overflows, it will drive an output RESET HIGH pulse at the RST pin.

Unit 14   Introduction to SIEMENS PLC

Product Overview

The S7-200 series of micro-programmable logic controllers (Micro PLCs) can control a wide variety of devices to support your automation needs.

The S7-200 monitors inputs and changes outputs as controlled by the user program, which can include Boolean logic, counting, timing, complex math operations, and communications with other intelligent devices. The compact design, flexible configuration, and powerful instruction set make the S7-200 a perfect solution for controlling a wide variety of applications.

What's New?

The new features of the SIMATIC S7-200 include the following. Table 14-1 shows the S7-200 CPUs that support these new features.

- S7-200 CPU models CPU 221, CPU 222, CPU 224, CPU 224XP, and CPU 226 to include the following features.

New CPU hardware support: option to turn off run mode edit to get more program memory. CPU 224XP supports onboard analog I/O and two communication ports. CPU 226 includes additional input filters and pulse catch.

-New memory cartridge support: S7-200 explorer browser utility, memory cartridge transfers, compares, and programming selections.

-STEP 7-Micro/WIN, version 4.0, a 32-bit programming software package for the S7-200 to include new and improved tools that support the latest CPU enhancements: PID Auto-Tuning Control Panel, PLCs built-in Position Control Wizard, Data Log Wizard, and Recipe Wizard.

New diagnostic tool: configuring diagnostic LED.

New instructions: Daylight Savings time (READ_RTCX and SET_RTCX), Interval Timers

(BITIM, CITIM), Clear Interrupt Event (CLR_EVNT), and Diagnostic LED (DIAG_LED).

POU and library enhancements: new string constants, and added indirect addressing support on more memory types, improved support of the USS library read and write parameterization for Siemens master drives.

Improved Data Block: Data Block Pages, and Data Block auto-increment.

Improved usability of STEP 7-Micro/WIN.

Table 14-1 S7-200 CPUs

S7-200 CPU	Order Number
CPU 221 DC/DC/DC 6 Inputs/4 Outputs	6ES7 211-0AA23-0XB0
CPU 221 AC/DC/Relay 6 Inputs/4 Relays	6ES7 211-0BA23-0XB0
CPU 222 DC/DC/DC 8 Inputs/6 Outputs	6ES7 212-1AB23-0XB0
CPU 222 AC/DC/Relay 8 Inputs/6 Relays	6ES7 212-1BB23-0XB0
CPU 224 DC/DC/DC 14 Inputs/10 Outputs	6ES7 214-1AD23-0XB0
CPU 224 AC/DC/Relay 14 Inputs/10 Relays	6ES7 214-1BD23-0XB0
CPU 224XP DC/DC/DC 14 Inputs/10 Outputs	6ES7 214-2AD23-0XB0
CPU 224XP AC/DC/Relay 14 Inputs/10 Relays	6ES7 214-2BD23-0XB0
CPU 226 DC/DC/DC 24 Inputs/16 Outputs	6ES7 216-2AD23-0XB0
CPU 226 AC/DC/Relay 24 Inputs/16 Relays	6ES7 216-2BD23-0XB0

S7-200 CPU

The S7-200 CPU combines a microprocessor, an integrated power supply, input circuits, and output circuits in a compact housing to create a powerful Micro PLC, as shown in Fig. 14-1. After you have downloaded your program, the S7-200 contains the logic required to monitor and control the input and output devices in your application.

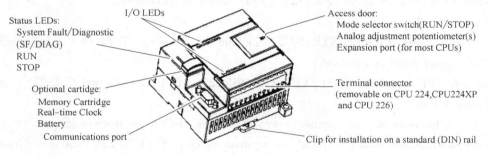

Fig. 14-1 S7-200 Micro PLC

Siemens provides different S7-200 CPU models with a diversity of features and capabilities that help you create effective solutions for your various applications.

S7-200 Expansion Modules

To better solve your application requirements, the S7-200 family includes a wide variety of expansion modules. You can use these expansion modules to add additional functionality to the S7-200 CPU, such as discrete modules, analog modules, intelligent modules, and other modules.

STEP 7—Micro/WIN Programming Package

The STEP 7-Micro/WIN programming package provides a user-friendly environment to develop, edit, and monitor the logic needed to control your application. STEP 7-Micro/WIN provides three program editors for convenience and efficiency in developing the control program for your application. To help you find the information you need, STEP 7-Micro/WIN provides an extensive online help system and a documentation CD that contains an electronic version of this manual, application tips, and other useful information.

Computer Requirements

STEP 7-Micro/WIN runs on either a personal computer or a Siemens programming device, such as a PG 760, as shown in Fig. 14-2. Your computer or programming device should meet the following minimum requirements.

- Operating system: Windows 2000, and Windows XP (Professional or Home).
- At least 100M bytes of free hard disk space.
- Mouse (recommended).

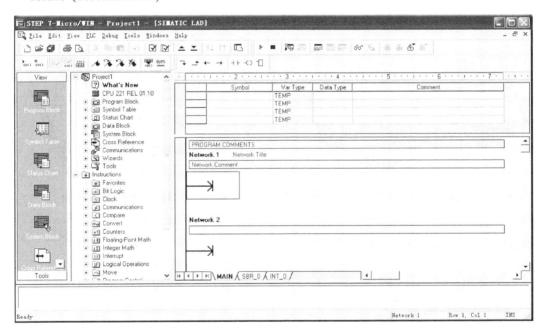

Fig. 14-2 STEP 7-Micro/WIN

Installing STEP 7-Micro/WIN

Insert the STEP 7-Micro/WIN CD into the CD-ROM drive of your computer. The installation wizard starts automatically and prompts you through the installation process.

Communications Options

Siemens provides two programming options for connecting your computer to the S7-200: a direct connection with a PPI Multi-Master cable, or a Communications Processor (CP) card with an MPI cable. The PPI Multi-Master programming cable is the most common and economical method of connecting your computer to the S7-200. This cable connects the communications port of the S7-200 to the serial communications of your computer. The PPI Multi-Master programming cable can also be used to connect other communications devices to the S7-200.

Display Panels

Text Display Unit (TD 200 and TD 200C)

The TD 200 and TD 200C are 2-line, 20-character text display devices that can be connected to the S7-200, as shown in Fig. 14-3. Using the TD 200 wizard, you can easily program the S7-200 to display text messages and other data pertaining to your application. The TD 200 and TD 200C provide a low cost interface to your application by allowing you to view, monitor, and change the process variables pertaining to your application.

TP070 and TP170 micro Touch Panel Displays

The TP070 and TP170 micro are touch panel display devices that can be connected to the S7-200, as shown in Fig. 14-4. This touch panel provides you with a means to customize your operator interface. These devices can display custom graphics, slider bars, application variables, custom user buttons, and so forth, by means of a user-friendly touch panel.

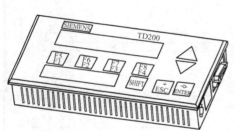

Fig. 14-3   Text Display Unit (TD 200 and TD 200C)

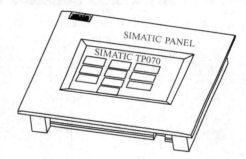

Fig. 14-4   Touch Panel Unit

New Words and Phrases

1. Boolean logic   布尔逻辑
2. complex math operations   复杂数学运算
3. input filters   输入滤波器
4. pulse catch   脉冲捕获
5. enhancement ［inˈhɑːnsmənt］ n. 增强；增加；提高；改善
6. built-in ［ˌbiltˈin］ adj. 嵌入的；内置的；固有的
7. recipe ［ˈresəpi］ n. 烹饪法；食谱；方法；秘诀
8. diagnostic ［ˌdaiəgˈnɔstik］ adj. 诊断的，判断的；特征的
9. parameterization ［ˌpærəˌmitəriˈzeiʃən］ n. 参数化，参数化法
10. Daylight Savings time   n. 夏令时
11. interval ［ˈintəvəl］ n. ［军事］间隔；间隔时间；（戏剧、电影或音乐会的）幕间休息

12. Interrupt Event   *n.* 中断事件
13. PPI (Point to point interface)   *n.* 点对点接
14. PID (proportional-integral-derivative)   比例积分微分
15. Auto-Tuning Control Panel   自动整定控制面板

Notes

1. The compact design, flexible configuration, and powerful instruction set make the S7-200 a perfect solution for controlling a wide variety of applications.

参考译文：紧凑的结构、灵活的配置和强大的指令集使 S7-200 成为各种控制应用的理想解决方案。

2. New and improved tools that support the latest CPU enhancements: PID Auto-Tuning Control Panel, PLCs built-in Position Control Wizard, Data Log Wizard, and Recipe Wizard.

参考译文：支持最新 CPU 增强功能的新软件工具和改进过的软件工具：PID 自动整定控制面板、PLC 内置位置控制向导、数据归档向导和配方向导。

3. New instructions: Daylight Savings time (READ_RTCX and SET_RTCX), Interval Timers (BITIM, CITIM), Clear Interrupt Event (CLR_EVNT), and Diagnostic LED (DIAG_LED).

参考译文：新指令：夏令时（READ_RTCX 和 SET_RTCX）、间隔定时器（BITIM, CITIM）、清除中断事件（CLR_EVNT）以及诊断 LED（DIAG_LED）。

Exercises

1. Answer the following questions according to the text.

a) What is the programming software package for S7-200?

b) What is the most common and economical method of connecting your computer to the S7-200?

c) How to install STEP 7-Micro/WIN?

2. Translate the following sentences into English.

数字量输入：在每个扫描周期的开始，CPU 会读取数字量输入的当前值，并将这些值写入过程映像输入寄存器。

3. Translate the following sentences into Chinese.

Analog inputs: The S7-200 does not update analog inputs from expansion modules as part of the normal scan cycle unless filtering of analog inputs is enabled. An analog filter is provided to allow you to have a more stable signal. You can enable the analog filter for each analog input point.

Unit 15   Technology of Hydraulic Pressure

The purpose of the pump is, of course, to give pressure to the oil; in other words, to give power to the machine. The purpose of the valves is to control the flow of oil apply the power when and where it may be needed. To illustrate as simply as possible how this is accomplished in a "circuit", that is, in the run of oil from the reservoir, through the pump, the valves, the driven unit, and back to the reservoir, references are made to diagrams shown in Fig. 15-1 and Fig. 15-2.

First, get the general idea of the circuit from Fig. 15-1 (omitting the feeding mechanism), then a clear understanding of the operation of the speed-control and reverse valves (Fig. 15-2), after which it should not be difficult to understand the details of Fig. 15-2 including the feed mechanism. The diagram in Fig. 15-1 shows the speed-control valve open (speed-control piston pulled out), permitting the exhaust through V port (9) to the reservoir. The machine is running; oil from the reservoir is being pumped in the direction of the arrows through R1 to the intake port (1) in the valve, out through (3) to the right-hand end of the cylinder, and forces the piston (and worktable) to the left. This pushes the oil on the left-hand side of the piston out of the cylinder and down through (4), across the spool through port (2), and on down through the V port (9) to the reservoir. The instant the valve is changed, the flow of oil through the valve is reversed (as shown in Fig. 15-2), and the piston travels in the opposite direction. Referring to Fig. 15-2a (which is an enlargement of the valve in Fig. 15-1), oil from the reservoir (and pump), through R1 to (1) to (3) to the cylinder, pushes the piston to the left; oil on the other side of the piston escapes from the cylinder down through (4) to (2) to (9) to R2 to the reservoir. Notice that the oil can enter (1) but cannot enter (5) because it is stopped by the land (11), also that it cannot go through (12) because it is stopped by the land (13).

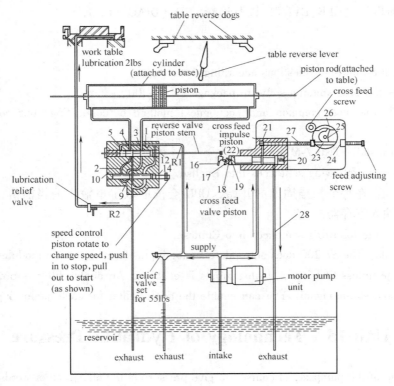

Fig. 15-1　A typical hydraulic circuit

In Fig. 15-2b, the valve is shown shifted to the left. This merely closes (1) and open (5). Oil flows now through R1 to (5) to (4) to the left side of the cylinder, and at the same time the oil on the right side of the cylinder exhausts through (3) to (2) to (9) to R2 to the reservoir. Note that, as in Fig. 15-2a, the oil can flow on as stated; elsewhere it is shut off by the lands on the valve plunger. Referring to speed-control plunger, the V port (9) is simply a notch cut in the side. Rotating the plunger a slight amount serves to reduce the size of the port, and of course, the amount of the oil that can pass through the port, and consequently the speed of the driven piston in the cylinder and therefore of the sliding worktable.

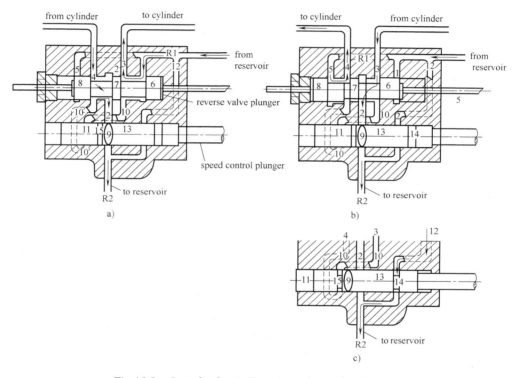

Fig. 15-2 Control valve in Fig. 15-1 enlarged for clearness

When the valve is pushed way in, as in Fig. 15-2c, no oil is discharged through the V port (9) and the table remains stationary, oil from the pump by-passing through the exhaust line (12) and the space (14). This is the way to stop the table, instead of shutting off (9) entirely, since it avoids forcing the oil through the relief valve. It will be noted in Fig. 15-2 that, with the plunger pushed way in and the power traverse stopped, the space (15) in the control plunger opens the line (10), and oil may flow from either end of the cylinder through (10), making hand feed of the table possible. That is as the table is fed back and forth by hand, the oil which fills the cylinder and the pipe line is pushed by the piston back and forth from one side to the piston to the other through the pipe line. To understand this more easily, refer first to Fig. 15-2c and note the line is open through (10) to (15); then refer to Fig. 15-2a, and imagine this line is open as in Fig. 15-2c. Then, as the table is moved from right to left, the oil will flow from the cylinder through (4) to (10) through (15) through the rest of (10) and on up through (3) to the cylinder. When the stroke is reversed and the table is moved left to right, the oil will be reversed and will flow (3) to (10) through the space (15) and up through the rest of (10) to (4), then to the left end of the cylinder.

Every fluid-power system used one or more pumps to pressurize the hydraulic fluid. The fluid under pressure, in return, performs work in the output section of the fluid-power system. Thus, the pressurized fluid may be used to move a piston in a cylinder or to turn the shaft of a hydraulic motor. The purpose of a pump in a fluid-power system uses low pressures (100 psi or less) to do work. Where a large work output is required, high pressure (10000 psi or more) may be used. So we find that every modern fluid-power system uses at least one pump to pressurize the fluid.

Three types of pumps are used in fluid-power systems, 1) rotary, 2) reciprocating, and 3) centrifugal pumps.

Simple hydraulic systems may use one type of pump. The tread is to use pumps with the most satisfactory characteristics for the specific tasks involved. In matching the characteristics of the pump to the requirements of the hydraulic system, it is not unusual to find two types of pumps in series. For example, a centrifugal pump may be used to supercharge a reciprocating pump, or a rotary pump may be used to supply pressurized oil for the controls associated with a reversing variable-displacement reciprocating pump.

These are built in many different designs and are extremely popular in modern fluid-power system. The most common rotary-pump designs used today are spur-gear, internal-gear generated rotor, sliding-vane, and screw pump. Each type has advantages that make it most suitable for a given application. Spur-gear pump. (Fig. 15-3) has two mating gears are turned in a closely fitted casing. Rotation of one gear, the driver, causes the second or follower gear to turn. The driving shaft is usually connected to the upper gear of the pump. When the pump is first started, rotation of gears forces air out the casing and into the discharge pipe. This removal of air from the pump casing produces a partial vacuum on the suction side of the pump. Fluid from an external reservoir is forced by atmospheric pressure into the pump inlet. Here, the fluid is trapped between the teeth of the upper and lower gears and the pump casing. Continued rotation of the gears forces the fluid out of the pump discharge.

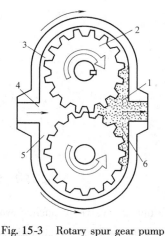

Fig. 15-3 Rotary spur gear pump
1—discharge (outlet)   2—drive gear
3—Oil is carried around housing in chambers formed between teeth
4—suction (inlet)
5—Vacuum is created here as teeth enmesh. Oil enters from reservoir
6—and forced out of pressure port as teeth go back into mesh

Pressure rise in a spur-gear pump is produced by the squeezing action on the fluid as it is expelled from between the meshing gear teeth and casing. A vacuum is formed in the cavity between the teeth as they enmesh, causing more fluid to be drawn into the pump. A spur-gear pump is a constant-displacement unit; its discharge is constant at a given shaft speed. The only way the quantity of fluid discharged by spur-gear pump of the type in

Fig. 15-3 can be regulated is by varying the shaft speed. Modern gear pumps used in fluid-power systems develop pressures up to about 3000 psi. These pumps have a number of vanes which are free to slide into or out of slots in the pump rotor. When the rotor is turned by the pump driver, centrifugal force, spring, or pressurized fluid causes the vanes to move outward in their sots and bear against the inner bore of the pump casing or against a cam ring. As the rotor revolves, fluid flows between the vanes when it passes the suction port. This fluid is carried around the pump casing until the discharge port is reached. Here, the fluid is forced out of the casing and into the discharge pipe.

Pressure control valves are used in hydraulic circuits to maintain desired pressure levels in various parts of the circuits. A pressurecontrol valve maintains the desired pressure level by (1) diverting higher-pressure fluid to a lower-pressure area, thereby limiting the pressure in the higher pressure area, or (2) restricting fluid to flow into another area. Valves that divert fluid can be safety, relief, counter-balance, sequence, and unloading types. Valves that restrict fluid to flow into another area can be of the reducing type. A pressurecontrol valve may also be defined as either a normally closed or a normally opened two-way valve. Relief, sequence, unloading, and counterbalance valves are normally closed, and two-way valves are partially or fully opened while performing their design function.

The application of hydraulic power to the operation of machine tools is by no means new, though its adoption on such a wide scale as exists at present, is comparatively recent. It was in fact that the development of the modern self-contained pump unit stimulated the growth of this form of machine tool operation. Machine tool hydraulic drive offers a great many advantages. One of them is that it can give infinitely-variable speed control over wide ranges. In addition, it can change the direction of drive as easily as it can vary the speed. The flexibility and resilience of hydraulic power is another great virtue of this form of drive. Apart from the smoothness of operation thus obtained, a great improvement is usually found in the surface finish on the work and the tool can make heavier cuts without detriment and will last considerable longer without regrinding. By far the greater proportion of machine tool hydraulic drives are confined to the linear motion, a rotary pump being used to actuate one or more linear hydraulic motors in the form of double-acting hydraulic rams, usually of the piston type. In some cases, as in certain hydraulic lathes both the linear motions of the cutting tool and the rotary motion of the work may be hydraulically driven and/or controlled. Such rotary motions are produced by the use of a rotary hydraulic motor.

New Words and Phrases

1. hydraulic [haiˈdrɔːlik] *adj.* 液压的
2. preference [ˈprefərəns] *n.* 优先选择
3. compact [kəmˈpækt] *adj.* 紧凑的，紧密的，简洁的
4. sleeplessly [ˈsliːplisli] *adj.* 无阶级的
5. valve [vælv] *n.* 阀
6. piston [ˈpistən] *n.* 活塞
7. chamber [ˈtʃeimbə] *n.* 油腔

Notes

1. Valves that restrict fluid to flow into another area can be of the reducing type.

参考译文：限制液体流到另一区域的阀类可能是减压型的。

2. A pressurecontrol valve may also be defined as either a normally closed or a normally opened two-way valve.

参考译文：压力控制阀也可被定义为常闭式或常开式二通阀。

3. It was in fact that the development of the modern self-contained pump unit stimulated the growth of this form of machine tool operation.

参考译文：事实上，正是现代整体泵的出现，促进了这种机床操作形式的发展。

Exercises

1. Put the following words or sentences into Chinese.

a) In hydraulic system, high torque at low speed is required; a very compact unit is needed; a smooth transmission, free of vibration, is required; easy control of speed and direction is necessary; or output speed must be varied steplessly.

b) In general terms, hydraulic drives may be divided into rotary and linear types. Rotary drives produce a rotating motion, whilst linear devices in the form of piston and cylinder units produce a reciprocating movement.

2. Translate the following sentences into English.

a) 一般来说，液压驱动可分为旋转运动方式和直线运动方式。

b) 当油液进入液压马达或油缸时，转化为机械能。

c) 液压马达和液压泵的结构几乎相同，任何泵都可用做马达。

Unit 16　Pneumatic Technology

Pneumatic technology is an engineering technology which using compressor as the power source, compressed air as medium for energy or signal transfer. Compressed air transports the energy to the actuator, output force (straight cylinder) or torque (a cylinder or air motor) through a series of control elements. Pneumatic technology is a science and technology which researches the flow law of compressed air.

Pneumatic technology is the best mean to implement low cost automation compared with hydraulic, electric technologies. Pneumatic element has a simple structure, compact, easy to manufacture, low price, medium inexhaustible, less pollution to the environment, high reliability, and long life, which can store the energy and can be transmitted over a long distance, because of the air compressibility.

Compared with machinery, hydraulic, and electric technologies, pneumatic technology has a wide work adaptability, easy to achieve rapid linear reciprocating motion, the swinging, high-speed mobile, and output force. It is conveniently to adjustment the movement speed, easy to change the direction of the movement, and it is to safe work and reliable in the harsh environment, easy to implement moisture proof, explosion-proof. Installation and control has a higher degree of freedom. It has the overload protection ability. Pneumatic technology also has certain disadvantages: compared with machinery, hydraulic, and electric technologies, difficult to reach the uniform state, because of the air compressibility, air cylinder movement speed vulnerable to the influence of the load, low speed movement, friction larger proportion of thrust, creep phenomenon easily, low speed poor stability, and the output force of the cylinder smaller.

Pneumatic system consists of the following components and devices components.

Air source device: the occurrence of compressed air and compressed air storage device, and purify the auxiliary device. It provides the high quality compressed air for the system. Execution element: the gas pressure can be converted into mechanical energy and complete the work of action of components, such as cylinder, air motor. The control elements: components which control gas pressure, flow, and direction motion, such as all kinds of valves; components which can complete certain logic function, such as pneumatic sensor and the equipment processing signal. Pneumatic auxiliary: auxiliary components of the pneumatic system, such as muffler, pipes, fittings, and so on.

Pneumatic technology, whose full name is pressure drive and control technology, is one of the most effective means of automatic production process and mechanization, having many advantages such as high speed and high efficient, safe, clean, easy maintenance, low cost, easy maintenance, and so on, widely used in light industry machinery field, also playing more and more important role in food packaging and production in the process.

Industrial robot is the most typical representative of the pneumatic technology application. It can instead of the human hand, wrist, and finger, can correct and quickly do grab or let go of such as the subtle movements. In addition to the application of industrial production, outside in the playground on a roller coaster of brake, mechanical production animal, and the interior of the human form chimes, adopted the pneumatic technology to realize the tiny movements.

Hydraulic pressure can take great output force but sensitivity is not enough; on the other hand, electricity driving objects always need to use some gears, and at the same time, we can't ignore the dangers of the leakage. And compared with this, using pneumatic technology not only safe but also no pollution, it can realize the tiny movement even in a small space. Its output force more than electric, even has the same size. In the production line, it can realize move forward, stop, turn to, and small simple act, and it can not lack in automation equipment. It is also applied in other aspects, such as manufacturing silicon wafer production line is indispensable in the resistance of the process of liquid daub use quantitative output pump and the combination of the surrounding machine.

In addition, although pneumatic technology has already got a wide range of applications in each industrial sector, but, there have many fairly big differences applications between them. For application of pneumatic technology, the most basic condition is to have an air compressor, so for other place with use of the air compressor, pneumatic technology application is much more convenient.

Especially in some non-productive processing departments, such as animal husbandry, the planting industry, or the clothing industry, the situation is even more so. In the machine equipment manufacturing, all have air compressors, and also pneumatic technology has been applied, each application project in essence similar, therefore, we can summarize the pneumatic technology used in machine equipment manufacture.

1. Agriculture

(1) Field operations

The field work equipment, the inclination, ascension and rotating device; crop protection and weed control equipment; bags hoist and other handling equipment.

(2) Animal breeding

Feed measurement and transmission; waste collection and cleared; eggs separation system; ventilation equipment; cutting the wool and slaughter.

(3) Animal feed production

The equipment of animal feed production and the equipment packaging of loading and unloading; measurement device; mixing and weighing system; devices which can storage, count, and monitoring.

(4) Forestry

Forestry is a production department to protect the ecological environment to maintain the balance, to cultivate and protect forest in order to obtain the wood and other forest products, to play the characteristics and use the natural forest protection.

(5) Planting

Greenhouse ventilation equipment; harvester; the equipment of fruits and vegetables classification.

2. Public facilities

(1) Thermal power plant

The ventilation equipment used in boiler room; Remote valve; Pneumatic switch control.

(2) Nuclear power plant

Fuel and absorber into device; the interlock between blend materials mouth and manual controlled air lock; Test and measurement device; Automatic operation system.

(3) Water supply system

Water level control; remote valve; sewage and waste water treatment of the rake gear mesh and control valve operation.

3. Mining

The auxiliary equipment used in the open air or in the underground mines, directly or indirectly, in mining mineral.

4. Chemical industry

Containers cover hermetic seal; measurement system; (mixture) mixing pole regulation; lab chemicals mixing device; immersed in the cell lifting devices; control valves operation; weighing device control; packing; die making; water level control and process control system.

5. The oil industry

Similar to the chemical industrial, the auxiliary equipment used in the factory and the laboratory.

6. Plastic rubber industry

(1) Plastic production

Mass transfer and distribute material control system; control valves operation system; material box 1 door.

(2) Plastics processing

Cut off the machine operation; modeling in closed, forming and welding machine; the closure device of the moulding.

(3) Rubber processing

Safety device; the transmission and production installations in the control and drive; mixer and curing the closed compressor device; test equipment.

New Words and Phrases

1. component [kəm'pəunənt] *n.* 组成部分，部件，元件
2. sensitivity [ˌsensi'tiviti] *n.* 敏感；敏感度
3. ventilation [ˌventi'leiʃən] *n.* 通风设备；通风方法

Notes

1. Pneumatic technology is a science and technology which researches compressed air flow law.
参考译文：气动技术是研究压缩空气流动规律的科学技术。

2. Compared with machinery, hydraulic, and electric technologies, pneumatic technology has a wide work adaptability, easy to achieve rapid linear reciprocating motion, the swinging, high-speed mobile, and output force, adjustment the movement speed conveniently, easy to change the direction of the movement.
参考译文：与机械、液压和电气技术相比，气动技术的工作适应性更广泛，易于实现快速的直线往复运动、摆动、高速移动和输出力，运动速度调节方便，改变运动方向简单。

3. Pneumatic technology, whose full name is pressure drive and control technology, is one of the most effective means of automatic production process and mechanization, having many advantages such as high speed and high efficient, safe, clean, easy maintenance, low cost, easy maintenance, and so on.
参考译文：气动技术，全称气压传动与控制技术，是生产过程自动化和机械化的最有效手段之一，具有高速高效、清洁安全、低成本、易维护等优点。

Exercises

1. Answer the following questions according to the text.

a) What does "low cost" (in the line 1, paragraph 2) refer to?

b) What does pneumatic technology used in machine equipment manufacture?

c) According to the text, what are the disadvantages of pneumatic technology?

2. Translate the following sentences into English.

a) 气压传动由于工作压力较低（一般在1MPa以下），且有可压缩性，传递动力小，所以运动不如液压传动平稳。

b) 气动系统是以气体（常用压缩空气）为工作介质传递动力或信号的系统。

Unit 17　Electrical Engineering

Electrical engineering (referred to as EE) is one of the core disciplines of modern science and technology, and also is the key discipline indispensable in the field of high and new technology. For example, it is because of the great progress of electronic technology to promote the computer network on the basis of the advent of the information age, and will change work mode of the human life. The development of electrical engineering with a potential prospects, it makes high employment rate of today's students.

In a sense, the development level of electrical engineering representative national scientific level and technological progress level. Because of this, the education and scientific research of electrical engineering have a very important position in the developed countries university occupies.

The universities in the United States, electrical engineering discipline called something different, in some schools called electrical engineering, some called the electrical engineering and information department, others called electrical engineering and computer science, and so on. This subject (department) is rather different from our electrical engineering discipline in scientific research, teaching, and academic organization form. It can adjust our subject direction, and improve the teaching and scientific research by learning foreign discipline state, teaching and scientific research direction.

The definition of traditional electrical engineering is the sum total used for creation of electric and electronic system of the subject. This definition has already been very broad, but with the rapid development of science and technology, the concept of electrical engineering has been far beyond the above definition of the category in the 21 century. The professor of Stanford University noted that: today, the electrical engineering covers almost all the electrons and photons of relevant engineering behavior. The huge growth of the width of domain knowledge, requests us to check even to construct the electrical engineering discipline direction, the curriculum and its content, so that electrical engineering discipline can effectively respond to the needs of their students, the needs of the community, the progress of science and technology, and dynamic research environment.

In the coming years, the biggest factors impact on the development of electrical engineering, including the following respects.

1. The decisive influence of information technology

Information technology is widely defined as comprehensively including computer, the computer network of the world high-speed broadband, and communication system, used to sense, process, store and display all kinds of information and related technical support. The information technology has a dominant influence to the development of electrical engineering. Information technology continues to increase exponentially, which depends to a great extent on the electrical engineering in the various subjects for technology innovation. In return, the progress of information technology provides the update more advanced tools basis for electrical engineering technical innovation.

2. Cross each other widened between the physical science

Due to the invention of the transistor and the development of the large scale integrated circuit manufacturing technology, the solid electronics play a great role of the growth of electrical engineering in the back of the 50 years of the 20 century. In the future, the key of electrical engineering discipline is the electrical engineering and physical sciences close contact between the cross, which will widen to biological systems, photonics, and microelectromechanical system (MEMS). In the 21 century, some important new devices, new systems, and new technologies will come from these fields.

3. Rapid change

Rapid progress of the technology and changes with each day of the analysis method and the design method, we must think or review the past solutions of the engineering problems to comprehensive every few years. This has big effect to hire new professors and train our students.

The power is the important material foundation for developing production and improving the level of human life. The application of the power in deepening and development, electrical automation is the important symbol of modernization in the national economy and people's life. At the international level, a long time in the future, the demand of electric power will continue to expand, and the demand of scientific and technical workers of electrical engineering and automation is on the rise. Electrical automation is used for industrial control systems to ensure the normal operation of equipment and production of qualified products. Modern industry is not full manual, relying on people to operate, but made by the machine. Starting machine, it can automatically run, which is electrical automation. Electrical automation uses relay, sensors, and electrical components to realize sequence control and time control of the process. Some of the other instruments or servo motors, can according to the change of external environment feedback to internal, change output, and achieve the purpose of stability.

4. Talents training target of electrical engineering

Electrical engineers trains broad caliber " composite" senior engineering and technical persons which can relate to electrical engineering system operation, automatic control, power electronic technology, information processing, test analysis, research and development, economic management, and the application of electronics and computer technology field.

5. Main courses of electrical engineering

Main courses of electrical engineering include the circuit principle, analog electronic technology, digital electronic technology, microcomputer principle and application, signal and system, automatic control principle, electrical motors and drives, power electronics technology, electric power drag automatic control system, electrical control technology and PLC application, microcomputer control technology, and power supply technology.

New Words and Phrases

1. category [ˈkætəgəri] *n.* 范畴
2. Stanford [ˈstænfəd] *n.* 斯坦福
3. curriculum [kəˈrikjuləm] *n.* 课程
4. comprehensive [ˌkɔmpriˈhensiv] *adj.* 综合的
5. microelectromechanical 微机电
6. stability [stəˈbiliti] *n.* 稳定性
7. electrical engineering 电气工程

Notes

1. It is because of the great progress of electronic technology to promote the computer network on the basis of the advent of the information age, and will change work mode of the human life.

参考译文：正是电子技术的巨大进步才推动了以计算机网络为基础的信息时代的到来，并将改变人类的生活工作方式。

2. Today, the electrical engineering covers almost all the electrons and photons of relevant engineering behavior.

参考译文：今天的电气工程涵盖了几乎所有与电子、光子有关的工程行为。

3. At the international level, a long time in the future, the demand of electric power will continue to growth, and the demand of scientific and technical workers of electrical engineering and automation is on the rise.

参考译文：就目前国际水平而言，在今后相当长的时期内，电力的需求将不断增长，电气工程及其自动化科技工作者的需求量呈上升态势。

Exercises

Translate the following sentences into English.

a) 了解国外学科状态及教学、科研方向，对调整我们的学科方向、提高教学、科研水平具有十分重要的作用。

b) 电气自动化就是利用继电器、感应器等电气元件实现顺序控制和时间控制。

c) 今后若干年内，对电气工程发展影响最大的主要因素，包括以下几个方面。

Unit 18　Introduction to NC

At the end of World War II, several events led to an experiment that changed metal manufacturing. In the early 1940s, the need to produce military products, such as airplanes, accelerated technical research. As a result, the products being produced at the end of the war were too complex in shape or closely tolerance for practical manufacturing. To assist in engineering calculations, a computer was developed at the University of Pennsylvania. The ENIAC (Electrical Numerical Integrator and Calculator) as it was called, was a huge mass of tubes and wires. It was difficult to program and very slow by today's standards, but it was a computer. Then, numerical control machines and computer were used to develop today's computer numerical control (CNC) machines. The Parsons Corporation, the United States Air Force, and the Massachusetts Institute of Technology each played a role in this development. In 1946, the Parsons Corporation tried to find accurate ways to make complicated aircraft parts. In an effort to generate rotor blade for a helicopter, they experimented with complicated tables of coordinates and manual machines. To generate the compound curves, they placed one human operator per axis handle, and called each move out in turn. This was slow and prone to errors. The Parsons Corporation then turned its attention to automatically generation these shapes. It seemed that an automatic method was possible. In 1949, John Parsons then set up a demonstration of his ideas for the Air Force. With a demonstration, Parsons convinced the Air Force to award a research contract. Shortly thereafter, Parsons set up a subcontract with the Servomechanisms Laboratory of the Massachusetts Institute of Technology (MIT). After three years, MIT built the first NC milling machine. In 1952, a vertical spindle milling machine ran the first true NC-produced parts, see Fig. 18-1. The electrical cabinet took up more floor space than the machine. However, it is the beginning that changes machining forever. The prototype numerical control machine developed by MIT used a punched tape to generate movements of three axes. This machine was capable of making curved shapes, quickly, accurately, and reliably. In 1960, machine manufacturers began to make NC equipment that many companies could afford, i.e. equipment was on the market at a price that allowed many shops to purchase their first NC machine.

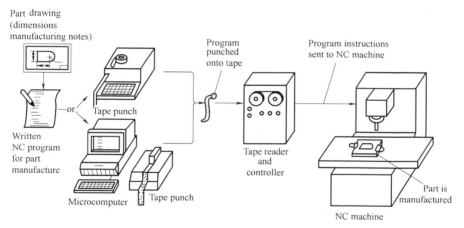

Fig. 18-1 Components of traditional NC system

Numerical control has been used in industry for over 50 years. Simply put, numerical control is a method of automatically operating a manufacturing machine based on a code of letters, numbers, and special characters. A complete set of coded instructions for executing an operation is called a program. The program is translated into corresponding electrical signals for input to motors which run the machine. Numerical control machines can be programmed manually. If a computer is used to create a program, the process is known as computer-aided programming. The approach taken in this text will be in the form of manual programming. Traditionally, numerical control systems have been composed of the following components.

Tape punch: converts written instruction into a corresponding hole pattern. The hole pattern is punched into tape which passes through this device. Much older units used a typewriter device called a flexowriter. Newer devices include a microcomputer coupled with a tape punch unit.

Tape reader: reads the hole pattern on the tape and converts the pattern to a corresponding electrical signal code.

Controller: receives the electrical signal code from the tape reader and subsequently causes the NC machine to respond.

NC machine: responds to programmed signals from the controller. Accordingly, the machine executes the required motions to manufacture a part (spindle rotation on, spindle rotation off, table or spindle movement along programmed axis directions, etc.), shown in Fig. 18-1.

NC systems offer some of the following advantages over manual methods of production.

1. Better control of tool motions under optimum cutting conditions.
2. Improved part quality and repeatability.
3. Reduced tooling costs, tool wear, and job setup time.
4. Reduced time to manufacture parts.
5. Reduced scrap.
6. Better production planning and placement of machining operations.

A computer numerical control (CNC) machine is an NC machine with the added feature of an on-board computer. The on-board computer is often referred to as the machine control unit or MCU. Control units for NC machines are usually hard wired. This means that the functions of a machine are controlled by the physical electronic elements that are built into the controller. The on-board computer, on the other hand, is "soft" wired. Thus, the functions of the machine are encoded into the computer at the time of manufacture. They will not be erased when the CNC machine is turned off. Computer memory that holds such information is known as ROM or read-only memory. The MCU usually has an alphanumeric keyboard for direct or manual data input (MDI) of part programs. Such programs are stored in RAM or the random-access memory portion of the computer. They can be played back, edited, and processed by the controller. All programs residing in RAM, however, are lost when the CNC machine is turned off. These programs can be saved on auxiliary storage devices such as punched tape, magnetic tape, or magnetic disk. Newer MCU units have graphics screens that can display not only the CNC program but also the cutter paths generated and any errors in the program.

The components found in many CNC systems are shown in Fig. 18-2.

Machine control unit: generates, stores, and processes CNC programs. The machine control unit also contains the machine motion controller in the form of an executive software program.

NC machine: responds to programmed signals from the machine control unit and manufactures the part.

An NC machine simply following directions from a program, as a slave, there is neither flexibility nor authority over the work being done, and the program is external to the control such as a

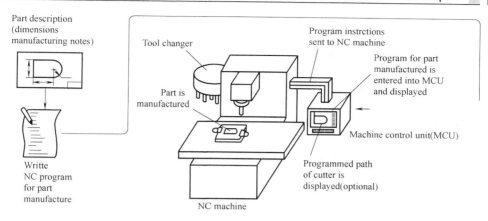

Fig. 18-2 Components of modern CNC system

punched tape. Fixed-all information must be in the program. Computer numerical control (CNC) means that data stored within the MCU memory direct the operation of a machine. The microprocessor allows the machine to exercise authority over the operation. A CNC controller can make calculations and decisions.

A CNC machine is more flexible than an NC machine. Similar to NC, a CNC machine follows program directions. In addition, it assumes responsibility for calculations and decisions, and contains a memory in which the program resides. This means that the program can be changed easily. The memory that can be edited is called a random access memory (RAM). A CNC machine can detect problems and help the operator to correct them. It can also communicate with the operator and external devices, such as robots and central programming computers.

Machining centers are the latest development in CNC technology. These systems come equipped with automatic tool changers with the capability of changing up to 90 or more tools. Many are also fitted with the movable rectangular worktables called pallets. The pallets are used to automatically load and unload work pieces. A single setup machining center can perform such operations as milling, drilling, tapping, boring, and so on. Additionally by utilizing indexing heads, some centers are capable of executing these tasks on many different faces of a part and at specified angles. Machining centers shown in Fig. 18-3 save production time and cost by reducing the need for moving a part from one machine to another.

Turning centers shown in Fig. 18-4 with increased capacity tool changers are also making strong appearance in modern production shops. These CNC machines are capable of executing many different types of lathe cutting operations simultaneously on rotating part.

In addition to machining centers and turning centers, CNC technology has also been applied to many other types of manufacturing equipment. Among these are wire electrical discharge machines (wire EDM) and laser cutting machines. Wire EDM machines utilize a very thin wire as an electrode. The wire is stretched between two rollers and cuts the part like a band-saw. Material is removed by the erosion caused by a spark that moves horizontally with the wire. CNC is used to control horizontal table movements.

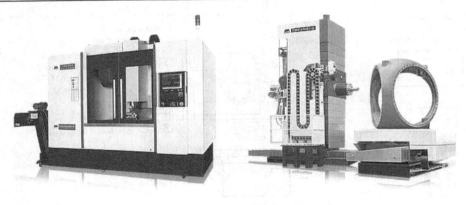

a) Vertical machining center   b) Horizontal machining center

Fig. 18-3　machining center

a) Vertical turning center   b) Horizontal turning center

Fig. 18-4　Turning center

Laser cutting CNC machines utilize an intense beam of focused laser light to cut the part. Material under the laser beam undergoes a rapid rise in temperature and is vaporized. If the beam power is high enough, it will penetrate through the material. Because no mechanical cutting forces are involved, laser cut parts have a minimum of distortion. These machines have been very effective in machining slots and drilling holes.

New Words and Phrases

1. military ['militəri] adj. 军事的，军用的
2. tolerance ['tɔlerəns] n. 公差，宽容，容许量；vt. 给（机器部件等）规定公差
3. Pennsylvania [pensil'veninjə] 宾夕法尼亚洲（美国州名）
4. servomechanism ['sə:vən'mekənizəm] n. 伺服系统
5. spindle ['spindl] n. 主轴，轴
6. machining centers 加工中心
7. turning centers 车削中心
8. milling ['miliŋ] n. 铣削

9. simultaneously [siməl'teiniəsly] *adv.* 同时发生地；同步地

Notes

1. In an effort to generate an accurate rotor blade for a helicopter, they experimented with complicated tables of coordinates and manual machines.

参考译文：为了精确制造出一个直升机的转子叶片，他们用具有多坐标的复杂工作台与手动机床进行实验。

2. NC systems offer some of the following advantages over manual methods of production.

参考译文：相对手工制造而言，NC 系统具有下列优势。

3. Reduced tooling cost, tool wear, and job setup time.

参考译文：降低加工成本、减少刀具磨损和作业准备时间。

4. NC machine: responds to programmed signals from the machine control unit and manufactures the part.

参考译文：数控机床：响应机床控制单元的程序信号并加工零件。

5. A CNC machine can detect problems and help the operator to correct them. It can also communicate the operator and external devices, such as robots and central programming computers.

参考译文：计算机数控机床能够检测错误，并帮助操作员改正错误，还能与操作员以及机器人、中央编程计算机等外设进行通信。

Exercises

1. Translate the following sentences into Chinese.

a. At the conclusion of this unit, you will be able to describe the history of numerical control.

b. Simply put, numerical control is a method of automatically operating a manufacturing machine based on a code of letters, numbers, and special characters.

c. A computer numerical control (CNC) machine is an NC machine with the added feature of an on-board computer.

2. Try to answer the following questions.

a. What components comprise a traditional NC machine?

b. What advantages does numerical control offer over manual methods?

c. Name at least three types of manufacturing equipment, which CNC technology has been applied?

Reading Material A

Power Saving Modes of Microcontrollers

Idle Mode

In the idle mode, the CPU puts itself to sleep while all the on-chip peripherals remain active. The mode is invoked by software. The content of the on-chip RAM and all the special function registers remain unchanged during this mode. The idle mode can be terminated by any enabled interrupt or by a hardware reset.

Note that when idle mode is terminated by a hardware reset, the device normally resumes program execution from where it left off, up to two machine cycles before the internal reset algorithm takes control. On-chip hardware inhibits access to internal RAM in this event, but access to the port pins is not inhibited. To eliminate the possibility of an unexpected write to a port pin when idle mode is terminated by a reset, the instruction following the one that invokes idle mode should not write to a port pin or to external memory.

Power-down Mode

In the Power-down mode, the oscillator is stopped, and the instruction that invokes Powerdown is the last instruction executed. The on-chip RAM and the special function registers retain their values until the Power-down mode is terminated. Exit from Power-down mode can be initiated either by a hardware reset or by activation of an enabled external interrupt (INT0 or INT1). Reset redefines the SFRs but does not change the on-chip RAM. The reset should not be activated before VCC is restored to its normal operating level and must be held active long enough to allow the oscillator to restart and stabilize.

New Words and Phrases

1. invoke [in'vəuk] vt. 祈求；提出或援引……以支持或证明；唤起；引起
2. terminate ['tə:mineit] vt. & v. 结束；使终结；解雇；到达终点站
3. resume [ri'zju:m] vt. 重新取得；再占有；取回
4. leave off vt. 停止（做）某事，戒掉
5. algorithm ['ælgəriðəm] n. 运算法则；演算法；计算程序
6. inhibit [in'hibit] v. 抑制；禁止
7. eliminate [i'limineit] vt. 排除；消除；除掉
8. stabilize ['steibəlaiz] vt. & v. （使）稳定，（使）稳固；使稳定平衡

Reading Material B

Guidelines for Designing a Micro PLC System

There are many methods for designing a micro PLC system. The following general guidelines can apply to many design projects. Of course, you must follow the directives of your own company's procedures and the accepted practices of your own training and location.

Partition Your Process or Machine

Divide your process or machine into sections that have a level of independence from each other. These partitions determine the boundaries between controllers and influence the functional description specifications and the assignment of resources.

Create the Functional Specifications

Write the descriptions of operation for each section of the process or machine. It includes the following topics: I/O points, functional description of the operation, states that must be achieved before allowing action for each actuator (such as solenoids, motors, and drives), description of the operator interface, and any interfaces with other sections of the process or machine.

Design the Safety Circuits

Identify equipment requiring hard-wired logic for safety. Control devices can fail in an unsafe manner, producing unexpected startup or change in the operation of machinery. Where unexpected or incorrect operations of the machinery could result in physical injury to people or significant property damage, consideration should be given to the use of electro-mechanical overrides which operate independently of the S7-200 to prevent unsafe operations.

Specify the Operator Stations

Based on the requirements of the functional specifications, drawings of the operator stations are created.

Create the Configuration Drawings

Based on the requirements of the functional specification, configuration drawings of the control equipment are created.

-Overview showing the location of each S7-200 in relation to the process or machine.

-Mechanical layout of the S7-200 and expansion I/O modules (including cabinets and other equipment).

-Electrical drawings for each S7-200 and expansion I/O module (including the device model numbers, communications addresses, and I/O addresses).

Create a List of Symbolic Names (optional)

If you choose to use symbolic names for addressing, a list of symbolic names for the absolute addresses are created. It include not only the physical I/O signals, but also the other elements to be used in your program.

New Words and Phrases

1. guideline ['gaidlain] *n.* 指导方针
2. directive [di'rektiv] *n.* 指令；<美>命令，训令，指令；方针
3. partition [pɑː'tiʃən] *vt.* 分开，隔开；区分；分割
4. independence [ˌindi'pendəns] *n.* 独立，自主
5. boundary ['baundəri] *n.* 分界线；范围；(球场)边线
6. assignment [ə'sainmənt] *n.* 分配；任务，工作，作业；指定，委派
7. actuator ['æktjueitə] *n.* 激励者；[电脑] 执行机构；[电](电磁铁) 螺线管；[机] 促动器
8. solenoids ['səulinɔid] *n.* 螺线管
9. identify [ai'dentifai] *vt.* 识别，认出；确定；使参与；把……看成一样
10. cabinet ['kæbinit] *n.* 内阁；柜橱；小房间；展览艺术品的小陈列室
11. symbolic [sim'bɔlik] *adj.* 象征的，象征性的

Reading Material C

Hydraulic Power Transmission

Hydraulic drives are used in preference to mechanical systems when powers is to be transmitted between point too far apart for chains or belts; high torque at low speed is required; a very compact unit is needed; a smooth transmission, free of vibration, is required; easy control of speed and direction is necessary; or output speed must be varied sleeplessly.

Electrically driven oil pressure pumps establish an oil flow for energy transmission, which is fed to hydraulic motor or hydraulic cylinder, converting it into mechanical energy. The control of the oil flow is by means of valves. The pressurized oil flow produces linear or rotary mechanical motion. The kinetic energy of oil flow is comparatively low, and therefore the term hydrostatic driver is sometimes used. There is little constructional difference between hydraulic motor and pumps. Any pump may be used as a motor. The quantity of oil flowing at any given time may be varied by means of regulating valves or the use of variable-delivery pumps. In general terms, hydraulic drives may be divided into rotary and linear types. Rotary drives produce a rotating motion whilst linear devices in the form of piston and cylinder unites produce a reciprocating movement.

All hydraulic motors function broadly is in accordance with the some basic principle. A pressurized fluid is alternately forced into and removed from a chamber. The filling cycle begins with minimum chamber volumes. When the chamber reaches its maximum volume (the maximum capacity), the filling is ended by isolating the chamber from the supply line. The oil is then returned to the oil pump through the return lines and at the same time the next chamber is filled with oil.

New Words and Phrases

1. oil pressure pump 油泵
2. hydraulic motor 液压马达
3. hydraulic cylinder 油缸
4. valve [vælv] n. 阀
5. regulating valve 调节阀
6. relief valve 安全阀
7. kinetic energy 动能
8. hydraulic driver 静压传动
9. variable-delivery pump 变量泵
10. chamber ['tʃeimbə] n. 油腔

Reading Material D

Tooling for Computer Numerical Control Machines

To the onlooker one of the most startling aspects of computer numerical control machine is the rapid metal-removal rates used. That there are cutting tools capable of withstanding such treatment can seem quite incredible. Add to this indexing times of less than one second and automatic tool changing providing a "chip-to-chip" time of around five seconds, it is easy to understand why many production engineers consider tooling to be the most fascination aspect of computer numerical control machine.

Although high-speed steel (HSS) is used for small-diameter drills, taps, reamers, end mills, and spot drills, the bulk of tooling for computer numerical control machine involves the use of cemented carbide.

The physical properties necessary in a cutting tool are hardness at the metal-cutting temperature, which can be as high as 600℃, and toughness. High-speed steel is tougher than cemented carbide but not as hard as it, therefore, cannot be used at such high rates of metal removal. On the other hand, the lack of toughness of cemented carbide presents problems, and this has meant that a tremendous amount of research has gone into developing carbide grades that, when adequately supported, are able to meet the requirements of modern machining techniques.

The hardness of cemented carbides is almost equal to that of diamond. It derives this hardness from its main constituent, tungsten carbide. In its pure form tungsten carbide is too brittle to be used as a cutting tool, so it is pulverized and mixed with cobalt. The mixture of tungsten carbide and cobalt powder is pressed into the required shape and then sintered. The cobalt melts and binds the tungsten carbide grains into a dense, nonporous structure.

In addition to tungsten carbide, other hard materials such as titanium and tantalum carbides are used, and by providing tungsten carbide tools with a thin layer of titanium carbide, resistance to wear and useful life are increased by up to five times.

The production of a part invariably involves the use of a variety of cutting tools, and the machine has to cater for their use. The way in which a range of cutting tools can be located and securely held in position is referred to as a tooling system and is usually an important feature of the machine tool manufacturers advertising literature.

The tooling system for a machining center is illustrated to make the use of tool holders with standard tapers, and has a feature that can be very helpful in keeping tooling costs to a minimum.

Automatic tool changer tool holders are multipurpose devices that are designed to meet the following needs.

1. Be easily manipulated by the tool changing mechanism.

2. Ensure repeatability of a tool. The tool is centered in the spindle such that the tool's relation to the work is repeated (within tolerance) every time the tool is used.

3. Provide fast and easy off-line tool assembly.

Many different types of mechanisms have been designed for storing and changing tools. The three most important ones are turret head, carousel storage with spindle changing, and matrix magazine storage with pivot insertion tool changer. Tool storage magazines may be horizontal or vertical.

The first type of system is found on older NC drilling machines. The tools are stored in the spindles of a device called a turret head. When a tool is called by the program, the turret rotates (indexes) it into position. The tool can be used immediately without having to be inserted into a spindle. Thus, turret head designs provide for very fast tool changes. The main disadvantage of turret head changers is the limit on the number of tool spindles that can be used.

Systems of the second type are usually found in vertical machining centers. Tools are stored in a coded drum called a carousel. The drum rotates to the space where the current tool is to be stored. It moves up and removes the current tool, then rotates the new tool into position and places it into the spindle. On larger systems, the spindle moves to the carousel during a tool change.

Chain-type storage matrix magazines have been popular in machining centers since early 1972. This type of system permits an operator to load many tools in a relatively small space. The chain may be located on the side or the top of the CNC machine. These positions enable tools to be stored away from the spindle and work. This will ensure a minimum of chip interference with the storage mechanism and a maximum of tool protection. Upon entering a programmed tool change, the system advices to the magazine and the old tool in the spindle. The magazine then advices to the space where the old tool is to be stored. The arm executes a rotation again and inserts the new tool into the spindle and the old tool into the magazine. A final rotation returns the arm back to its parked position.

Two methods of tool identification are currently in use. One is the bar code designation. The code is imprinted and fastened to the tool. When the program calls for a specific tool, the controller looks for a particular tool code, not a specific location. Another tool identification system uses a computer microchip, which is part of the tool or tool holder. The microchip contains the tool identification number and information related to the parameters of the tool. A special sensor reads these information and transfers them to the machine controller.

New Words and Phrases

1. onlooker ['ɔn‚lukə, 'ɔːn-] *n.* 旁观者
2. cemented carbide 硬质合金钢
3. tremendous [tri'mendəs] *adj.* 极大的，巨大的
4. tungsten carbide 碳化钨，硬化合金
5. sinter ['sintə] *vt.* 使烧结
6. variety [və'raiəti] *n.* 变化多样性，种种，品种，种类
7. multipurpose ['mʌlti'pəːpəs] *adj.* 多种用途的，多目标的
8. mechanism ['mekənizəm] *n.* 机械装置，机构，机制
9. repeatability *n.* 可重复性，反复性，再现性

Part V　Applications of Engineering

Unit 19　Introduction to WinCC Flexible

1. Introduction to SIMATIC HMI

Introduction

Maximum transparency is essential for the operator who works in an environment where processes are becoming more complex, and requirements for machine and plant functionality are increasing. The Human Machine Interface (HMI) provides this transparency.

The HMI system represents the interface between man (operator) and process (machine/plant). The PLC is the actual unit which controls the process. Hence, there are an interface between the operator and WinCC flexible (at the HMI device) and another interface between WinCC flexible and the PLC. An HMI system assumes the following tasks.

- Process visualization

The process is visualized on the HMI device. The screen on the HMI device is dynamically updated based on process transitions.

- Operator control of the process

The operator can control the process by means of the GUI. For example, the operator can preset reference values for the controls or start a motor.

- Displaying alarms

Critical process states automatically trigger an alarm. For example, when the setpoint value is exceeded.

- Archiving process values and alarms

The HMI system can log alarms and process values. This feature allows you to log process sequences and to retrieve previous production data.

- Process values and alarms logging

The HMI system can output alarms and process value reports. This allows you to print out production data at the end of a shift, for example.

- Process and machine parameter management

The HMI system can store the parameters of processes and machines in recipes. For example, you can download these parameters in one pass from the HMI device to the PLC to change over the product version for production.

SIMATIC HMI

SIMATIC HMI offers a totally integrated, single-source system for manifold operator control and monitoring tasks. With SIMATIC HMI, you always master the process and always keep your machinery and units running.

Examples of simple SIMATIC HMI systems are small touch panels for use at machine level.

SIMATIC HMI systems used for controlling and monitoring production plants represent the upper end of the performance spectrum. These include high-performance client/server systems.

Integration of SIMATIC WinCC flexible

WinCC flexible is the HMI software for future-proof machine-oriented automation concepts with comfortable and highly efficient engineering. WinCC flexible combines the following benefits.

- Straightforward handling
- Transparency
- Flexibility

2. WinCC flexible system overview

(1) Components of WinCC flexible

WinCC flexible engineering system

The WinCC flexible engineering system is the software for handling all your essential configuring tasks. The WinCC flexible edition determines which HMI devices in the SIMATIC HMI spectrum can be configured.

WinCC flexible runtime system

WinCC flexible runtime system is your software for process visualization. You execute the project in process mode in runtime system.

WinCC flexible options

The WinCC flexible options allow you to expand the standard functionality of WinCC flexible. A separate license is needed for each option.

(2) WinCC flexible engineering system

Introduction

WinCC flexible is an engineering system for all your configuring tasks. WinCC flexible has a modular design. With each higher edition, you expand the spectrum of supported devices and WinCC flexible functionality. You can always migrate projects to a higher edition by means of a PowerPack.

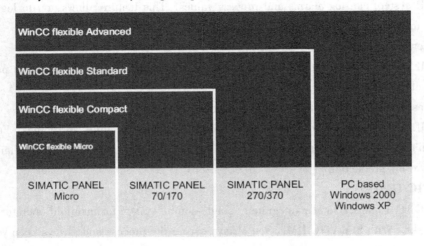

WinCC flexible covers a performance spectrum ranging from Micro Panels to simple PC visualization. The WinCC flexible functionality is thus comparable with that of products of the ProTool family and the TP Designer. You can integrate your existing ProTool projects into WinCC flexible.

New Words and Phrases

1. transparency [træns'pærənsi] n. 透明；透明度；透明性；透明的东西
2. operator ['ɔpəreitə] n. （机器、设备等的）操作员，机务员
3. plant [plɑ:nt] n. 设备；工厂
4. functionality [ˌfʌŋkəʃə'næliti] n. 功能；功能性；设计目的
5. HMI (Human Machine Interface) 人机界面
6. interface ['intəfeis] n. 界面；<计>接口；交界面
7. visualization [ˌvizjuəlai'zeiʃən] n. 形象（化），形象化，想象
8. dynamically [dai'næmikəli] adv. [计]动态地
9. transition [træn'ziʃən] n. 过渡，转变，变迁；[语]转换；[乐]变调
10. GUI (Graphical user interface) n. 图形用户界面
11. critical ['kritikəl] adj. 危急的；决定性的；[物]临界的
12. archive ['ɑ:kaiv] v. 存档；n. 档案文件，档案室
13. log [lɔg] vt. 把……记入航海日志；航行（……距离）
14. retrieve [ri'tri:v] vt. 取回；恢复；[计]检索；重新得到
15. manifold ['mænəˌfəuld] adj. 多种多样的；多方面的；有多种形式的；有多种用途的
16. spectrum ['spektrəm] n. 范围；系列，范围，幅度

Notes

1. Critical process states automatically trigger an alarm, for example, when the setpoint value is exceeded.

参考译文：过程的临界状态会自动触发报警，例如，当超出设定值时。

2. The HMI system can log alarms and process values. This feature allows you to log process sequences and to retrieve previous production data.

参考译文：HMI 系统可以记录报警和过程值。该功能使您可以记录过程序列，并检索以前的生产数据。

3. WinCC flexible is the HMI software for future-proof machine-oriented automation concepts with comfortable and highly efficient engineering.

参考译文：WinCC flexible 是一种前瞻性的面向机器的自动化概念的 HMI 软件，它具有舒适而高效的设计。

Exercises

1. Put the following into Chinese by reference to the text.

process visualization; setpoint value; straightforward handling; flexibility; range from

2. Please introduce the tasks about an HMI system.

3. Translate the following sentences into Chinese.

a) An HMI device which is directly connected to the PLC via the process bus is referred to as a single-user system. Single-user systems are generally used near production, but can also be deployed to operate and monitor independent part processes or system sections.

b) PLC with several HMI devices: several HMI devices are connected to one or more PLCs via a process bus (e.g. PROFIBUS or Ethernet). Such systems are deployed, for example, in a production line to operate the plant from several points.

c) HMI system with centralized functions: an HMI system is connected to a PC via Ethernet. The upstream PC assumes central functions, e.g. recipe management. The necessary recipe data records are provided by the subordinate HMI system.

Unit 20 Flexible Manufacturing Systems

A flexible manufacturing system integrates all major elements of manufacturing into a highly automated system. First utilized in the late 1960s, FMS consists of a number of manufacturing cells, each containing an industrial robot (serving several CNC machines) and an automated material-handling system, all interfaced with a central computer. Different computer instructions for the manufacturing process can be downloaded for each successive part passing through the workstation.

This system is highly automated and is capable of optimizing each step of the total manufacturing operation. These steps may involve one or more processes and operations (such as machining, grinding, cutting, forming, powder metallurgy, heat treating, and finishing), as well as handling of raw materials, inspection, and assembly. The most common applications of FMS to date have been in machining and assemble operations. A variety of FMS technology is available from machine-tool manufactures.

Flexible manufacturing systems represent the highest level of efficiency, sophistication, and productivity that has been achieved in manufacturing plants. The flexibility of FMS is such that it can handle a variety of part configurations and produce them in any order.

FMS can be regarded as a system which combines the benefits of two other systems: (a) the highly productive but inflexible transfer lines, and (b) job-shop production, which can produce large product variety on stand-alone machines, but inefficient.

Elements of FMS

The basic elements of a flexible manufacturing system are (a) workstations, (b) automated handling and transport of materials and parts, and (c) control system. The workstations are arranged to yield the greatest efficiency in production, with an orderly flow of materials, parts, and products through the system.

The types of machines in workstations depend on the type of production. For machining operations, they usually consist of a variety of three- to five-axis machining centers, CNC lathes, milling machines, drill processes, and grinders. Also various other equipment is included, such as that for automated inspection, assembly, and cleaning.

Other types of operations suitable for FMS include sheet metal forming, punching and shearing, and forging; they incorporate furnaces, forging machines, trimming presses, heat-treating facilities, and cleaning equipment.

Because of the flexibility of FMS, material-handling, storage, and retrieval systems are very important. Material handling is controlled by a central computer and performed by automated guided vehicles, conveyors, and various transfer mechanisms. The system is capable of transporting raw materials, blanks, and parts in various stages of completion to any machine (in random order) and at any time. Prismatic parts are usually moved on specially designed pallets. Parts having rotational symmetry (such as those for turning operations) are usually moved by mechanical devices and robots.

The computer control system of FMS is its brains and includes various software and hardware. This sub-system controls the machinery and equipment in workstations and the transporting of raw materials, blanks, and parts in various stages of completion from machine to machine. It also stores data and provides communication terminals that display the data visually.

Scheduling

Because FMS involves a major capital investment, efficient machine utilization is essential: machines must not stand idle. Consequently, proper scheduling and process planning are crucial.

Scheduling for FMS is dynamic, unlike that in job shops, where a relatively rigid schedule is followed to perform a set of operations. The scheduling system for FMS specifies the types of operations to be performed on each part, and it identifies the machines or manufacturing cells to be used. Dynamic scheduling is capable of responding to quick changes in product type and so is responsive to real-time decisions.

Because of the flexibility in FMS, no setup time is wasted in switching between manufacturing operations; the system is capable of performing different operations in different orders and on different machines. However, the characteristics, performance, and reliability of each unit in the system must be checked, to ensure that parts moving from workstation to workstation are of acceptable quality and dimensional accuracy.

Economic Justification of FMS

FMS installations are very capital-intensive, typically starting at well over $1 million. Consequently, a thorough cost-benefit analysis must be conducted before a final decision is made. This analysis should include such factors as the cost of capital, energy, materials, and labor, the expected markets for the products to be manufactured, and any anticipated fluctuations in market demand and product type. An additional factor is the time and effort required for installing and debugging the system.

Typically, an FMS system can be taken two to five years to install and at least six months to debug. Although FMS requires few, if any, machine operators, the personnel in charge of the total operation must be trained and highly skilled. These personnel include manufacturing engineers, computer programmers, and maintenance engineers.

Compared to conventional manufacturing system, some benefits of FMS are the following.

(1) Parts can be produced randomly, in batch sizes as small as one, and at lower unit cost.

(2) Direct labor and inventories are reduced, to yield major saving over conventional system.

(3) The lead times required for product changes are shorter.

(4) Production is more reliable, because the system is self-correcting, and product quality is uniform.

New Words and Phrases

1. Flexible Manufacturing Systems (FMS)　柔性制造系统
2. CNC (Computer Numerical Control)　计算机数控
3. integrate ['intigreit] vt. 集成，使一体化，积分　v. 结合
4. material-handling　物料输送，原材料处理
5. successive [sək'sesiv] adj. 继承的，连续的
6. workstation ['wəːksteiʃən] n. 工作站
7. grinding ['graindiŋ] n. 磨削
8. cutting ['kʌtiŋ] n. 切削
9. powder metallurgy　粉末冶金
10. finishing ['finiʃiŋ] n. 带式磨光，饰面，表面修饰，擦光
11. machine-tool　机床
12. sophistication [sə,fisti'keiʃən] n. 老练，成熟，精致，世故
13. incorporate [in'kɔːpəreit] adj. 合并的，结社的；v. 合并，组成公司
14. milling machine　铣床
15. drill press　n. 钻床，立式钻床
16. grinder ['graində] n. 磨床，研磨机，磨工
17. punch [pʌntʃ] v. 凿孔，冲板，冲压
18. shear ['ʃiəriŋ] v. 剪切
19. forging machine　锻造机
20. trimming press　冲拔罐修边机
21. retrieval system　回收系统
22. transfer mechanism　传输机械装置
23. pallet ['pælit] n. 托盘，货盘
24. terminal ['təːminəl] n. 终点站，终端，接线端
25. capital-intensive [,kæpitlin'tensiv] n. 资金集约型

Notes

1. These steps may involve one or more processes and operations (such as machining, grinding, cutting, forming, powder metallurgy, heat treating, and finishing), as well as handling of raw materials, inspection, and assembly.

参考译文：这些步骤可能包括一个或多个程序和操作（比如加工、磨削、切削、成型、

粉末冶金、热处理和修整），还有原材料的处理、检查和装配。

2. For machining operations, they usually consist of a variety of three-to five-axis machining centers, CNC lathes, milling machines, drill processes, and grinders.

参考译文：对于机械加工操作，它们由多个三轴或五轴加工中心、CNC 车削、铣削加工、钻孔和磨削组成。

3. Other types of operations suitable for FMS include sheet metal forming, punching and shearing, and forging; they incorporate furnaces, forging machines, trimming presses, heat-treating facilities, and cleaning equipment.

参考译文：其他适合 FMS 的操作包括板材成型、冲压和裁剪、锻造。它们把熔炉、锻造机器、裁剪、热处理设备和清洗设备融于一体。

4. Because of the flexibility of FMS, material-handling, storage, and retrieval systems are very important. Material handling is controlled by a central computer and performed by automated guided vehicles, conveyors, and various transfer mechanisms.

参考译文：由于柔性制造系统的柔性，物料处理、存储和回收等系统显得很重要。物料处理由一台中央计算机控制并由自动引导车、传送器和不同的传送机制执行。

5. Dynamic scheduling is capable of responding to quick changes in product type and so is responsive to real-time decisions.

参考译文：动态的计划安排能够对产品类型的迅速变化作出反应，并且对实时决策反应也是灵敏的。

6. However, the characteristics, performance, and reliability of each unit in the system must be checked, to ensure that parts moving from workstation to workstation are of acceptable quality and dimensional accuracy.

参考译文：但是，必须对系统中的每个制造单元的特点、性能和可靠性进行检验，以确保在工作站之间流动的工件满足质量和尺寸精度方面的要求。

7. Direct labor and inventories are reduced, to yield major saving over conventional system.

参考译文：直接劳工成本和库存减少，比起常规系统节约了许多。

Exercises

1. Put the following into Chinese by reference to the text.
manufacturing cells; industrial robot; automated material-handling system;
job-shop production; capital-intensive.

2. Please introduce the elements of FMS.

3. Translate the following sentences into Chinese.

a) The FMS is fundamentally an automated, conveyorized, computerized job shop. The system is complex to schedule. Because the machining time for different parts varies greatly, the FMS is difficult to link to an integrated system and often remains an island of expensive automation.

b) The development of FMSs began in the United States in the 1960s. The idea was to combine the high reliability and productivity of the transfer line with the programmable flexibility of the NC machine in order to be able to produce a variety of parts.

c) A FMS generally needs about three or four workers per shift to load and unload parts, change tools, and perform general maintenance. The workers in the FMS are usually high skilled and trained in NC and CNC.

Unit 21　CAD/CAM/CAPP

CAD is the abbreviation for Computer Aided Design, with the Chinese meaning of computer aided design, and refers to the computer as the auxiliary means to complete the whole process of product design. Generalized CAD includes the design and analysis of two aspects. The typical CAD hardware includes computer, one or more of the graphic display terminals, keyboard, and other peripherals. CAD software is responsible for system by computer graphics processing procedure plus some of the engineering function becomes easy user of the application. Examples of these applications, include parts of the stress-strain analysis, the dynamic response mechanism, the heat transfer calculation, and the numerical control part programming. From a user company to the next user company, the applications set may vary, because their production line, the manufacturing process, and customer market are different. These factors have increased CAD system of different requirements.

CAM is the abbreviation for Computer Aided Manufacturing, with the Chinese meaning of computer aided manufacturing. Generalized CAM is manufacturing process through direct or indirect contact of the computer and production equipment, with planning, design, management and control of product. CAM mainly includes the use of the computer to complete the NC programming, machining process simulation, NC machining, quality inspection, product assembly, debugging, and so on. The core of the CAM is computer numerical control (hereinafter referred to as the numerical control). The computer is used in the manufacturing process, or system. The Massachusetts Institute of Technology in 1952 first developed NC milling machine. Numerical control is characterized by the program instructions coded in a punched paper tape to control machine tools. Since then, a series of numerical control machine tools were invented to come out, including one called "machining center" multifunctional machine tool. It can preform from the knife warehouse automatic tool change and automatic conversion work position, and can continuously complete the sharp, drilling, tapping, dumplings, and other processes, which are all through the program instructions to control operation. Just changing the program instructions can change the processing. The processing flexibility is called the "soft" of numerical control.

CAPP is the abbreviation for Computer Aided Process Planning, with the Chinese meaning of computer aided process planning, and is the use of the technology of computer software and hardware environment, the use of computer numerical simulation, the logical judgment and reasoning, and other functions, to develop machining process of mechanical parts. Operator inputs original data of the components, processing conditions, and processing requirements to the computer, and the computer can automatically encode, until finally output through the optimization process card.

This work needs to be complex planned by engineers with rich experience in the production, and to be achieved with the help of computer graphics, database, and expert system of computer science and technology. Computer Aided Process Planning (CAPP) often acts as a bridge between the Computer Aided Design (CAD) and Computer Aided Manufacturing (CAM).

The research and development of CAPP technology is from the 1960s. In 1969, Norway launched the world's first CAPP system AUTOPROS, and realized commercialization in 1973. The United States of America in the end of 1960s and the beginning of 1970s began to develop CAPP system. In 1976, the International Computer-Aided Manufacturing-International (CAM-I) Incorporation of United States launched the most famous, the most widely used CAPP system, which is a milestone in the history of CAPP development. Thereafter, many CAPP systems came out. Shanghai Tongji University in 1982 developed the first CAPP system of China, TOJICAP.

CAPP system consists of five modules: obtaining information of parts, process decision, process database and knowledge base, interface, and technology file management and output.

According to the working principle, CAPP system can be divided into retrieval type, derivative type, and automatic generation type. Retrieval type CAPP system is used in the standard process. You can number the standard process of well designed parts, and then store them in a computer in advance. When you want to make the process of parts, you can search for, according to the input of part information, to find the right standard process. According to "similar parts have similar process" principle, through the retrieval a process of similar typical parts, again by adding or editing, the derivative type can create the process of a new parts. Automatic generation type and derivative type are not the same. According to the input of part information, this type relies on engineering data and decision-making method of the system, automatically generating the technological process of components.

CAPP has many advantages, such as making process design personnel free from the tedious and repetitive work, spending more time and energy to engage in more creative work; can greatly shorten the production cycle and the cycle of process design, for enterprises to improve the ability of rapidly changing market demand for quick response, and to improve the competitive ability of enterprise products in the market; contribute to summarize and inherit the valuable experience of process design; be conducive to the optimization and standardization of process design, to improve the quality of process design, to improve productivity, and to reduce manufacturing cost; create conditions for realizing enterprise information integration, in order to achieve concurrent engineering, agile manufacturing, and other advanced production mode.

By means of a CAPP system, we can solve the low efficiency of manual process design, poor consistency, unstable quality, and not easy to achieve optimization. There is important significance either of the single-piece and small-batch production or of mass production.

Reviewing of the development of CAPP, the research and application of CAPP always need around two aspects of requirement. One is the continuous improvement of its application, the other is constantly meeting the new requirement of new technology and manufacturing mode. Therefore, the development of CAPP, will be in the application scope, the depth and level of the application. The specific performance of the development trend are the following three aspects: the first one is the

CAPP system of an product life cycle, the second one is the CAPP system based on knowledge. and the third one is the reconfigurable CAPP system based on the platform of the technology.

New Words and Phrases

1. abbreviation [əˌbriːviˈeiʃən] n. 缩写字，缩写式；缩写，省略；缩短
2. with the help of   在……的帮助下，借助
3. achieve [əˈtʃiːv] vt. 完成，实现；达到，赢得
4. automatically [ˌɔːtəˈmætikəli] adv. 自动地；无意识地，不自觉地，机械地
5. commercialization [kəˌməːʃəlaiˈzeiʃən] n. 商品化，商业化
6. module [ˈmɔdjuːl] n. 组件，单元；（航天器的）舱
7. management [ˈmænidʒmənt] n. 管理，经营，处理；管理部门；经理部
8. competitive [kəmˈpetitiv] adj. 竞争的；竞争性的；好竞争的
9. integration [ˌintiˈgreiʃən] n. 整合，完成；集成
10. reconfigurable [riˈkɔnfigərəbl] adj. 可重构的；重新组态

Notes

1. Generalized CAM is manufacturing process through direct or indirect contact of the computer and production equipment, with planning, design, management and control of product. CAM mainly includes the use of the computer to complete the NC programming, machining process simulation, NC machining, quality inspection, product assembly, debugging, and so on.

参考译文：广义的 CAM 是指通过计算机与生产设备直接的或间接的联系，进行规划、设计、管理和控制产品的生产制造过程。主要包括使用计算机来完成数控编程、加工过程仿真、数控加工、质量检验、产品装配和调试等工作。

2. The research and development of CAPP technology is from the 1960s. In 1969, Norway launched the world's first CAPP system AUTOPROS, and realized commercialization in 1973.

参考译文：CAPP 技术的研究和发展源于 20 世纪 60 年代。1969 年挪威推出了世界上第一个 CAPP 系统 AUTOPROS，并于 1973 年商品化。

3. CAPP system consists of five modules: obtaining information of parts, process decision, process database and knowledge base, interface, and technology file management and output.

参考译文：CAPP 系统由五大模块组成：零件信息的获取、工艺决策、工艺数据库/知识库、人机界面和工艺文件管理/输出。

4. Retrieval type CAPP system is used in the standard process. You can number well designed parts standard process, and then store them in a computer in advance. When you want to make the process of parts, you can search for according to the input of part information, to find the right standard process.

参考译文：检索式 CAPP 系统适用于标准工艺，你可以事先对设计好的零件标准工艺进行编号，然后将它预存在计算机中，当你想制定零件的工艺过程时，可根据输入的零件信息进行搜索，查找合适的标准工艺。

Exercises

1. Answer the following questions according to the text.
a) Whether CAD, CAM, CAPP all need computer to assist finish?
b) Does Computer Aided Process Planning (CAPP) often act as a bridge between the Computer Aided Design (CAD) and Computer Aided Manufacturing (CAM)?
c) Which country rolled out the first CAPP system in the world?
d) Please describe the advantages of CAPP.

2. Put the following into Chinese by reference to the text.
Design   Indirect   Simulation   Product assembly   Determine   Logic
Processing   Programming   Database   Modular   Standardization   Mass production

3. Translate the following sentences into English.
a) 借助于CAPP系统，可以解决手工工艺设计效率低、一致性差、质量不稳定、不易达到优化等问题。
b) 上海同济大学在1982年开发出我国第一个CAPP系统。

Unit 22 Datasheet of IRB 140 Industrial Robot

IRB 140 industrial robot is an ABB robot. Its main applications include arc welding, assembly, cleaning/spraying, machine tending, material handling, packing, deburring, and so on.

Small, Powerful and Fast

Compact, powerful IRB 140 industrial robot is a six axis multipurpose robot that handles payload of 6kg, with long reach (810mm), as shown in Table 22-1. The IRB 140 can be floor mounted, inverted, or on the wall in any angle. Available as Standard, Foundry Plus 2, Clean Room, and Wash versions, all mechanical arms completely IP67 protected, making IRB 140 easy to integrate in and suitable for a variety of applications. Uniquely extended radius of working area due to bend-back mechanism of upper arm, axis 1 has a rotation of 360 degrees even as wall mounted.

The compact, robust design with integrated cabling adds to overall flexibility. The Collision Detection option with full path retraction makes robot reliable and safe.

Using IRB 140T, cycle-times are considerably reduced where axis 1 and 2 predominantly are used. Reductions between 15-20% are possible using pure axis 1 and 2 movements. This faster versions are well suited for packing applications and guided operations together with PickMaster.

IRB 140 Foundry Plus 2 and Wash versions are suitable for operating in extreme foundry environments and other harsh environments with high requirements on corrosion resistance and tightness. In addition to the IP67 protection, excellent surface treatment makes the robot high pressure steam washable. The white-finish Clean Room version meets Clean Room Class 10 regulations, making it especially suitable for environments with stringent cleanliness standards.

Table 22-1　Specification and performance parameters of IRB 140 Series

IRB140 Specification			
Robot versions	Handling capacity	Reach of 5th axis	Remarks
IRB 140/IRB 140T	6kg	810mm	
IRB 140F/IRB 140TF	6kg	810mm	Foundry Plus 2 Protection
IRB 140CR/IRB 140TCR	6kg	810mm	Clean Room
IRB 140W/IRB 140TW	6kg	810mm	SteamWash Protection
Supplementary load (on upper arm and wrist)			
on upper arm		1kg	
on wrist		0.5kg	
Number of axes			
Robot manipulator		6	
External devices		6	
Integrated signal supply		12 signals on upper arm	
Integrated air supply		Max. 8 bar on upper arm	
Performance			
Position repeatability		0.03mm (average result from ISO test)	

Axis movement	Axis	Working range
	1	360°
	2	200°
	3	280°
	4	Unlimited (400° default)
	5	240°
	6	Unlimited (800° default)
Max. TCP velocity		2.5m/s
Max. TCP acceleration		20m/s^2
Acceleration time 0—1m/s		0.15s

Velocity		
Axis no.	**IRB 140**	**IRB 140T**
1	200°/s	250°/s
2	200°/s	250°/s
3	260°/s	260°/s
4	360°/s	360°/s
5	360°/s	360°/s
6	450°/s	450°/s

Cycle time		
5kg Picking side	IRB 140	IRB 140T
Cycle 25×300×25mm^3	0.85s	0.77s

(续)

Electrical Connections

Supply voltage	200~600V, 50/60Hz
Rated power	
Transformer rating	4.5kV·A
Power consumption typically	0.4kW

Physical

Robot mounting	Any angle
Dimensions	
Robot base	$400 \times 450 \text{mm}^2$
Robot controller H × W × D	$950 \times 800 \times 620 \text{mm}^3$
Weight	
Robot manipulator	98kg

Environment

Ambient temperature for	
Robot manipulator	5~45℃
Relative humidity	Max. 95%
Degree of protection	
Robot manipulator	IP67
Options	Foundry Plus 2
	SteamWash (high pressure steam washable)
	Clean Room, class 6 (certified by IPA)
Noise level	Max. 70dB (A)
Safety	Double circuits with supervision, emergency stops and safety functions 3-position enable device
Emission	EMC/EMI-shielded

Data and dimensions may be changed without notice.

Working range

Working range is shown in Fig. 22-1.

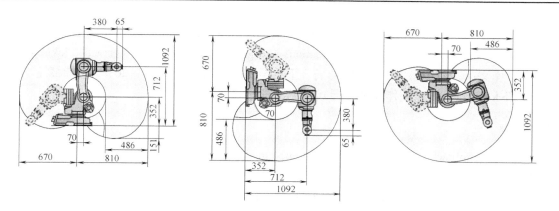

Fig. 22-1　Working Range of IRB 140

New Words and Phrases

1. arc welding （电）弧焊
2. spraying ['spreiiŋ] *n.* 喷雾
3. deburring [di'bə:riŋ] *n.* 修边，除去毛刺
4. radius ['reidjəs] *n.* 半径（距离）；用半径度量的圆形面积；半径范围
5. mount [maunt] *v.* 增加；上升 *vt.* 安装，架置；镶嵌，嵌入；准备上演
6. upper arm 上臂
7. robust [rəu'bʌst] *adj.* 精力充沛的；坚定的；粗野的，粗鲁的；需要体力的
8. collision [kə'liʒən] *n.* 碰撞；冲突；（意见，看法）的抵触
9. retraction [ri'trækʃən] *n.* 收回
10. predominantly [pri'dɔminəntli] *adv.* 占主导地位地；显著地；占优势地
11. reduction [ri'dʌkʃən] *n.* 减少；降低
12. corrosion resistance 耐（腐）蚀性，耐蚀力，抗腐（蚀）性
13. tightness ['taitnis] *n.* 坚固，紧密
14. stringent ['strindʒənt] *adj.* 严格的；追切的；（货币）紧缩的
15. emission [i'miʃən] *n.* 排放，辐射；排放物，散发物；（书刊）发行，发布（通知）
16. shield [ʃi:ld] *vt.* 保护；掩护；庇护；给……加防护罩

Notes

1. IRB 140 industrial robot is an ABB robot. Its main applications include arc welding, assembly, cleaning/spraying, machine tending, material handling, packing, deburring, and so on.

参考译文：IRB 140 是 ABB 公司的一款工业机器人，它的主要应用领域包括弧焊、装配、清理/喷雾、上下料、物料搬运、包装和去毛刺等。

2. Uniquely extended radius of working area due to bend-back mechanism of upper arm, axis 1 has a rotation of 360 degrees even as wall mounted.

参考译文：IRB 140 上臂采用后翻转机构，即使采用挂壁安装，第 1 轴仍可旋转 360 度，工作半径显著扩大。

3. Reductions between 15-20 % are possible using pure axis 1 and 2 movements. These faster versions are well suitable for packing applications and guided operations together with PickMaster.

参考译文：在使用第 1、第 2 轴的场合下，节拍时间可缩短 15% ~ 20%。这款高速型产品配套 PickMaster，是包装作业和引导式作业的理想之选。

Exercises

1. Put the following into Chinese by reference to the text.

floor mounted; Foundry Plus 2; Clean Room and Wash versions; Collision Detection; corrosion resistance and tightness; EMC/EMI-shielded

2. Translate the following sentences into Chinese.

a) Current research in robotics tends to be in two main areas, artificial intelligence and the related field of machine vision. A typical artificial intelligence problem in robotics might be to find a clear path for a robot through a cluttered environment. However, most of the research in this area is towards making intelligent machines. This is pure artificial intelligence and has little to do with robotics, except that any successes would have immediate applications to robots.

b) To make robots quicker, it is necessary to make the links lighter. This means they will be more flexible. In this book, we assume the links are perfectly rigid, an approximation of course. There is much current work on robots with significant flexibility. This is particularly important for robots in outer space, where weight is at a premium. It is also important to reduce the vibrations caused by this elasticity for accurate work on earth.

Unit 23 Machine Vision

Introduction

Machine Vision (MV) is concerned with the engineering of integrated mechanical-optical-electronic-software systems for examining raw materials, industrial equipment and manufacturing processes, in order to detect defects and improve quality, operating efficiency and the safety of both products and processes. It is also used to monitor and control machines used in manufacturing.

MV system may be divided into three categories according to their functions, visual recognition systems (VRS), visual inspection system (VIS), and visual guidance system (VGS). In practice, systems may be a combination of two or three categories, for example VRS + VIS, VRS + VGS, VGS + VIS, or VRS + VIS + VGS.

The emphasis of the VRS is to identify objects in a given area. Typical applications are vehicle recognition systems, object recognition by robots with vision function for the disabled and in the fields of medicine, arts, environmental monitoring, and aviation.

VIS is commonly used in inspection applications for detecting deformations, deviations from specifications, or missing elements. Such systems have an emphasis on checking the size and shape of objects. The significance is that they have the ability to carry out 100 percent on-line inspection and measurement. Many successful application examples can be found in industry. For example, the 100 percent visual inspection system of valve-stem seals, 3D mechanical parts visual inspection, and automatic PCB inspection.

VGS is concerned with the position of an object and directs a controller to move the object. VGS has a measuring loop, which supplies position and/or orientations of objects. All highly automated industries have a potential for applying VGS, for example, raw parts feeding and car screen insertion systems in Rover.

The diagram of an archetypal MV system is shown in Fig. 23-1, where its multi-disciplinary nature is apparent. It emphases the need to integrate a variety of different technologies.

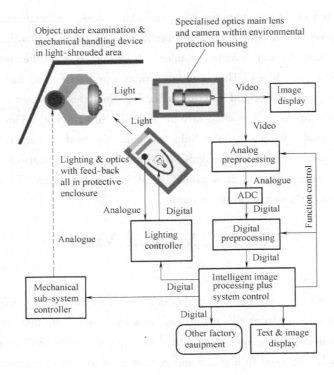

Fig. 23-1 Archetypal MV system

Elements of a Machine Vision System

Machine Vision necessarily involves the harmonious integration of elements of the following areas of study.

- Mechanical handling
- Lighting
- Optics (conventional, fibre optics, lasers, diffractive optics)
- Image sensor (camera or scanner)
- Electronics (digital, analogue and video)
- Signal processing
- Image processing
- Digital systems architecture
- Software
- Industrial engineering
- Human-computer interfacing
- Control systems
- Manufacturing

Industrial Applications of Machine Vision

The principal industrial applications of MV are inspection, robot guidance, process monitoring and control. We shall discuss each of these in turn.

Inspection

The word "inspection" is used in two ways: to refer to the specific task of detecting faults such as material defects, incomplete machining (e. g. untapped hole, missing chamfer, no final polishing, no slot on a screw, etc.), mislabelling, and in the more general sense, to include a range of applications such as counting, grading, sorting, locating, identifying, recognizing, etc.

It should be understood that there is no clear distinction between Automated Visual Inspection (AVI) and Robot Vision, since many "inspection" tasks require parts manipulation, and many object-handling applications also require identification and verification.

Robot Vision

The term "robot" can be applied to any machine that provides the ability to move its end effector (e. g. a gripper, or camera), under computer control, to a given point in 2-D or 3-D space. It may also be possible to control the orientation of the end effector. According to this definition, the following machines qualify to be called robots: numerically controlled milling machine, lathe, drill, electronic-component insertion machine, graph plotter, serpentine robotic arm, autonomously guided vehicle, etc.

The first industrial robots were "blind slaves", and proved to be useful in repetitive tasks, where the form and posture of the work-piece are predictable. However, they were severely limited in their total inability to cope with new and unexpected situations. While the positional accuracy of the robot may be very high, it may be expected to operate in an environment where the parts delivered to it are in an unknown position or posture, or may have a variable size or form. Vision is, of course, ideal for locating widgets, prior to directing a robot to pick them up. Without appropriate sensing of its environment, a robot can be dangerous; it can damage itself, other equipment, the widget and injure personnel nearby. Vision provides one of the prime sensing methods, for obvious reasons. A visually guided robot can "look ahead" and determine whether it is safe to move, before committing itself to what would be a dangerous manoeuvre. An intelligent visually guided robot can, for example, pick up an article that has fallen at random onto a table, even if it has not seen the article before. Without the use of vision, that would be impossible.

Process monitoring and visual control

Visual feed-back/feed-forward offers opportunities for process control that are not possible using any other sensing technology. It is possible to place the camera so that it observes either the product (partially or completely finished), the manufacturing process, or even waste products.

Refering to Fig. 23-2, the camera is observing the product. Notice that, in addition to the line controlling the accept/reject mechanism, there are two other outputs from the image-processing subsystem.

- *Feed-back* refers to adjust the operating parameters of a manufacturing process based on a machine located *up-stream*.
- *Feed-forward* refers to adjust the operating parameters of a manufacturing process based on a machine located *down-stream*.

In this case, the camera is placed between separate stages of processing so that it observes the

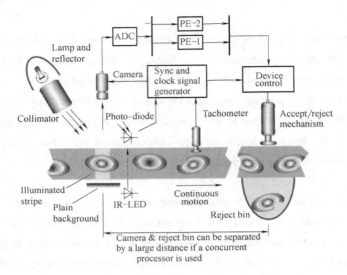

Fig. 23-2　The line controlling

partially-made product. The product might be either discrete piece-parts, or in the form of a continuous strip.

Concluding Remarks

Machine Vision systems are being used in manufacturing industry, in a very wide range of situations. To date, the electronics industry has been the most enthusiastic user of Machine Vision, but with strong support also coming from automobile, glass, plastics, aircraft, printing, pharmaceutical, food, medical products, etc. The technology has already shown itself to be capable of contributing to a very wide range of applications.

New Words and Phrases

1. integrated ['intəˌgretid] adj. 综合的，完整的
2. recognition [ˌrekəg'niʃən] n. 识别，认得
3. inspection [in'spekʃən] n. 检查，探伤，验收；目测；验证，校验
4. guidance ['gaidəns] n. 导航，指引；领导；制导，向导
5. carry out　完成，实现，贯彻，执行
6. valve-stem seals　阀杆密封
7. multi-disciplinary　多学科
8. human-computer interfacing　人机界面
9. end effector　末端执行器
10. serpentine robotic arm　蜿蜒（蛇行）机器手臂

Notes

1. The term "robot" can be applied to any machine that provides the ability to move its end effector (e.g. a gripper, or camera), under computer control, to a given point in 2-D or 3-D space.

It may also be possible to control the orientation of the end effector.

参考译文:"机器人"是指那些在计算机的控制下,能够移动末端执行器(比如夹具、摄像机等)到给定的二维或三维空间中的点的机械,机器人也可以控制末端执行器的方位。

2. Visual feed-back/feed-forward offers opportunities for process control that are not possible using any other sensing technology. It is possible to place the camera so that it observes either the product (partially or completely finished), the manufacturing process, or even waste products.

参考译文:视觉反馈/前馈为那些不能使用其他传感器技术的过程控制提供了机会,如通过安放摄像机检测产品(半成品或成品)、制造过程甚至废品。

Exercises

1. Answer the following questions according to the text.

a) According to their function, how many categories may Machine Vision systems be divided into?

b) Give some application examples of VRS, VIS, and their combination.

c) What is Machine Vision system constitute?

2. Put the following into Chinese by reference to the text.

visual recognition systems; visual inspection system; visual guidance system;

image sensor; signal processing; industrial engineering; defects;

AVI; 2-D; repetitive; prior to; intelligent visually guided robot.

3. Translate the following sentences into Chinese.

a) MV system may be divided into three categories according to their functions, visual recognition systems (VRS), visual inspection system (VIS), and visual guidance system (VGS).

b) The word "inspection" is used in two ways: to refer to the specific task of detecting faults such as material defects, incomplete machining (e.g. untapped hole, missing chamfer, no final polishing, no slot on a screw, etc.), mislabelling, and in the more general sense, to include a range of applications such as counting, grading, sorting, locating, identifying, recognising, etc.

c) Visual feed-back/feed-forward offers opportunities for process control that are not possible using any other sensing technology. It is possible to place the camera so that it observes either the product (partially or completely finished), the manufacturing process, or even waste products.

d) To date, the electronics industry has been the most enthusiastic user of Machine Vision, but with strong support also coming from automobile, glass, plastics, aircraft, printing, pharmaceutical, food, medical products, etc.

Unit 24 Automatic Assembly

Product Overview

The increasing need for finished goods in large quantities has, in the past, led engineers to search for and develop new methods of production. Many individual developments in the various branches of manufacturing technology have been made and have allowed the increased production of improved finished goods at lower cost. One of the most important manufacturing processes is the assembly process. This process is required when two or more component parts are brought together to produce the finished product.

The early history of the development of assembly process is closely related to the history of the development of mass-production methods. Thus, the pioneers of mass-production are also the pioneers of the modern assembly process. Their new ideas and concepts have brought significant improvements in the assembly methods employed on large-volume production.

However, although some branches of manufacturing engineering, such as metal cutting and metal forming process, have recently been developing very rapidly, the technology of the basic assembly process has failed to keep pace. Table 24-1 shows that in the United States the percentage of the total labor force involved in the assembly process varies from about 20% for the manufacture of farm machinery to almost 60% for the manufacture of telephone and telegraph equipment. Because of this, assembly costs often account for more than 50% of the total manufacturing costs. Statistical surveys show that these figures are increasing every year.

Table 24-1 Percentage of Production Workers Involved in Assembly

Industry	Percentage of workers involved in assembly
Motor vehicles	45.6
Aircraft	25.6
Telephone and telegraph	58.9
Farm machinery	20.1
Household refrigerators and freezers	32.0
Typewriters	35.9
Household cooking equipment	38.1
Motorcycles bicycles and parts	26.3

In the past few years, certain efforts have been made to reduce assembly costs by application of automation and modern techniques, such as ultrasonic welding and die-casting. However, success has been very limited and many assembly operators are still using the same basic tools as those employed at the time of the Industrial Revolution.

In the early days of manufacturing technology, the complete assembly of a product was carried out by a single operator and usually, this operator also manufactured the individual component part of the assembly. Consequently, it was necessary for the operator to be an expert in all the various aspects of the work, and training a new operator was a long and expensive task. The scale of production was often limited by the availability of trained operator rather than by the demand for the product.

In 1798, the United States needed a large supply of muskets and federal arsenals could not meet the demand. Because the war with the French was imminent, it was also not possible to obtain additional supplies from Europe. However, Eli Whitney, now recognized as one of the pioneers of mass production, offered to contract to make 10000 muskets in 28 months. Although it took 10 years to complete the contract, Whitney's novel ideas on mass production had been successfully proved. The factory at New Haven, Connecticut, built specially for the manufacture of the muskets, contained machines for producing interchangeable parts. These machines reduced the skills required by the various operators and allowed significant increases in the rate of production. In a historic demonstration in 1801, Whitney surprised his distinguished visitors when he assembled musket locks after randomly selecting parts from a heap.

The result of Eli Whitney's work brought about three primary developments in manufacturing methods. First, parts were manufactured on machines, resulting in a consistently higher quality than that of hand-made parts. These parts were now interchangeable and as a consequence assembly work was simplified. Second, the accuracy of the final product could be maintained at a higher standard, and third, production rates could be significantly increased.

Oliver Evans's conception of conveying material from one place to another without manual effort led eventually to further developments in automation for assembly. In 1793, he used three types of conveyors in an automatic flour mill, which required only two operators. The first operator poured grain into a hopper and the second one filled sacks with flour produced by the mill. All the intermediate operations were carried out automatically with conveyors carrying the material from operation to operation.

The next significant contribution to the development of assembly methods was made by Elihu Root. In 1849, Elihu Root joined the company that was producing Colt "six-shooters". Even though at that time the various operations of assembling the component parts were quite simple, he divided these operations into basic units that could be completed more quickly and with less chance of error. Root's division of operations gave rise to the concept "divide the work and multiply the output". Using this principle, assembly work was reduced to very basic operations and with only short periods of operator training, high efficiencies could be obtained.

Frederick Winslow Taylor was probably the first person to introduce the methods of time and motion study to manufacturing technology. The objective was to save the operator's time and energy by making sure that the work and all things associated to the work were placed in the best positions for carrying out the required tasks. Taylor also discovered that any work has an optimum speed of working which, if exceeds, results in a reduction in overall performance.

Undoubtedly, the principal contributor to the development of production and assembly methods was Henry Ford. He described his principles of assembly in the following word:

"First, place the tools and then men in the sequence of the operations so that each part shall travel the least distance while in the process of finishing".

"Second, use work slides or some other form of carrier so that when a workman complete his operation he drops the part always in the same place which must always be the most convenient place to his hand and if possible have gravity carry the part to the next workman".

"Third, use sliding assembly lines by which parts to be assembled are delivered at convenient intervals, spaces make it easier to work on them".

These principles were gradually applied in the production of the Model T Ford automobile.

New Words and Phrases

1. account for　占；计算出；解释，说明
2. ultrasonic ['ʌltrə'sɔnik] *n.* 超声波；*adj.* 超声波的，超声的
3. musket ['mʌskit] *n.* 火枪
4. arsenal ['ɑːsənəl] *n.* 兵工厂，军械库
5. imminent ['iminʃnt] *adj.* 危急的，急迫的
6. interchangeable [intə'tʃeindʒəbl] *adj.* 可互换的，通用的
7. distinguished [dis'tiŋgwiʃt] *adj.* 卓越的，杰出的
8. hopper ['hɔpə] *n.* 漏斗，料斗

Notes

1. Table 24-1 shows that in the United States the percentage of the total labor force involved in the assembly process varies from about 20% for the manufacture of farm machinery to almost 60% for the manufacture of telephone and telegraph equipment.

参考译文：表 24-1 显示出在美国，装配工艺中所使用的劳动力占总劳动力的百分比从农业机械制造业中的约 20% 到电话和电信设备制造业中的几乎 60% 的变化。

2. The scale of production was often limited by the availability of trained operator rather than by the demand for the product.

参考译文：生产规模不是受限于产品的需求，而是常常受限于可使用的训练有素的操作者的人数。

3. Second, use work slides or some other form of carrier so that when a workman complete his operation he drops the part always in the same place which must always be the most convenient place to his hand and if possible have gravity carry the part to the next workman.

参考译文：其次，使用工件滑道或其他形式的输送装置，以使得当一个工人完成其操作时能一直在其最顺手的同一位置放下零件，并且如果可能的话，让重力将零件送至下一个工人。

Exercises

1. Answer the following questions according to the text.

a) What does "these figures" (in the last line, paragraph 3) refer to?

b) According to the text, who was the first man to produce interchangeable parts in U. S. ?

c) Who is the principal contributor to the development of production and assembly methods?

d) Describe Henry Fords principle of assembly.

2. Put the following into Chinese by reference to the text.

finished; machinery; die-casting; a large supply of;

musket; arsenal; randomly; as a consequence.

3. Translate the following sentences into English.

a) 在制造工业中采用自动装配线可以提高生产效率，并能使产品维持在较高的质量

水平。

b）采用自动装配工艺必须先保证零件具有可互换性。

Reading Material A

Selecting the Communications Protocol for Your Network

The following information is an overview of the protocols supported by the S7-200 CPUs.
- Point-to-Point Interface (PPI)
- Multi-Point Interface (MPI)
- PROFIBUS

Based on the Open System Interconnection (OSI) seven-layer model of communications architecture, these protocols are implemented on a token ring network which conforms to the PROFIBUS standard as defined in the European Standard EN 50170. These protocols are asynchronous, character-based protocols with one start bit, eight data bits, even parity, and one stop bit. Communications frames depend upon special start and stop characters, source and destination station addresses, frame length, and a checksum for data integrity. The protocols can run on a network simultaneously without interfering with each other, as long as the baud rate is the same for each protocol.

Ethernet is also available for the S7-200 CPU with expansion modules CP243-1 and CP243-1 IT.

PPI Protocol

PPI is a master-slave protocol: the master devices send requests to the slave devices, and the slave devices respond, as shown in Fig. V A-1. Slave devices do not initiate messages, but wait until a master sends them a request or polls them for a response.

Masters communicate to slaves by means of a shared connection which is managed by the PPI protocol. PPI does not limit the number of masters that can communicate with any slave; however, you cannot install more than 32 masters on the network.

S7-200 CPUs can act as master devices while they are in RUN mode, if you enable PPI master mode in the user program. After enabling PPI master mode, you can use the

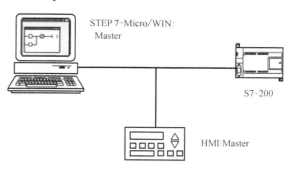

Fig. V A-1　PPI Network

Network Read or the Network Write instructions to read from or write to other S7-200s. While the S7-200 is acting as a PPI master, it still responds as a slave to requests from other masters.

PPI Advanced allows network devices to establish a logical connection between the devices. With PPI Advanced, there are a limited number of connections supplied by each device, as shown in Table V A-1 for the number of connections supported by the S7-200.

All S7-200 CPUs support both PPI and PPI Advanced protocols, while PPI Advanced is the only PPI protocol supported by the EM277 module.

Table ⅤA-1 Number of Connections for the S7-200 CPU and EM 277 Module

Module	BaudRate	Connections
S7-200CPU Port0	9.6k baud, 19.2k baud, or 187.5k baud	4
Port1	9.6k baud, 19.2k baud, or 187.5k baud	4
EM277 module	9.6k baud to 12M baud	6 per module

MPI Protocol

MPI allows both master-master and master-slave communications, as shown in Fig. ⅤA-2. To communicate with an S7-200 CPU, STEP 7-Micro/WIN establishes a master-slave connection. MPI protocol does not communicate with an S7-200 CPU operating as a master.

Network devices communicate by means of separate connections (managed by the MPI protocol) between any two devices. Communication between devices is limited to the number of connections supported by the S7-200 CPU or EM 277 module. The number of connections supported by the S7-200 is shown in Table ⅤA-1.

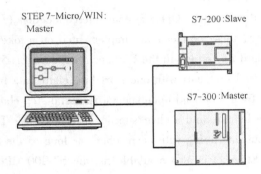

Fig. ⅤA-2 MPI Network

For MPI protocol, the S7-300 and S7-400 PLCs use the XGET and XPUT instructions to read and write data to the S7-200 CPU. For information about these instructions, refer to your S7-300 or S7-400 programming manual.

PROFIBUS Protocol

The PROFIBUS protocol is designed for high-speed communications with distributed I/O devices (remote I/O). There are many PROFIBUS devices available from a variety of manufacturers. These devices range from simple input or output modules to motor controllers and PLCs.

PROFIBUS networks typically have one master and several slave I/O devices, as shown in Fig. ⅤA-3. The master device is configured to know what types of I/O slaves are connected and at what addresses.

The master initializes the network and verifies that the slave devices on the network match the configuration. The master continuously writes output data to the slaves and reads input data from them.

When a DP master configures a slave device

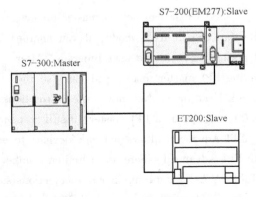

Fig. ⅤA-3 PROFIBUS Network

successfully, it then owns that slave device. If there is a second master device on the network, it has very limited access to the slaves owned by the first master.

TCP/IP Protocol

The S7-200 can support TCP/IP Ethernet communication through the use of an Ethernet (CP 243-1) or Internet (CP 243-1 IT) expansion module. Table V A-2 shows the baud rate and number of connections supported by these modules.

Table V A-2 Number of Connections for the Ethernet (CP243-1)
and the Internet (CP243-1 IT) Modules

Module	BaudRate	Connections
Ethernet (CP243-1) module	10 to 100M baud	8 general purpose connections
Internet (CP243-1 IT) module		1 STEP 7-Micro/WIN connection

Refer to the SIMATIC NET CP243-1 Communications Processor for Industrial Ethernet Manual or the SIMATIC NET CP243-1 IT Communications Processor for Industrial Ethernet and Information Technology Manual for additional information.

New Words and Phrases

1. protocol ['prəutə‚kɔːl] n. 礼仪；（数据传递的）协议；科学实验报告（或计划）
2. PROFIBUS abbr. (process field bus) 过程现场总线
3. Open System Interconnection 开放系统互连
4. token ring network 令牌环网
5. asynchronous [ei'siŋkrənəs] adj. 异步的
6. even parity 偶校验
7. frame [freim] n. 框架；眼镜框；组织；边框
8. checksum ['tʃeksʌm] n. 校验和
9. integrity [in'tegriti] n. 正直，诚实；完整；[计算机] 保存；健全
10. master-slave protocol 主从协议

Reading Material B

RobotStudio Overview

Offline programming using the Virtual Robot Technology is just like having the real robot on your PC!

Offline programming is the best way to maximize return on investment for robot systems. ABB's simulation and offline programming software, RobotStudio, allows robot programming to be done on a PC in the office without shutting down production. It also enables robot programs to be prepared in advance, increasing overall productivity.

RobotStudio provides the tools to increase the profitability of your robot system by letting you perform tasks such as training, programming, and optimization without disturbing production. This

provides numerous benefits including:
- Risk reduction
- Quicker start-up
- Shorter change-over
- Increased productivity

RobotStudio is built on the ABB Virtual Controller, an exact copy of the real software that runs your robots in production. It thus allows very realistic simulations to be performed, using real robot programs and configuration files identical to those used on the shop floor.

New Words and Phrases

1. Virtual Robot Technology   虚拟机器人技术

Reading Material C

Computer-Integrated Manufacturing System (CIMS)

A CIM system is commonly thought of as an integrated system that encompasses all the activities in the production system from the planning and design of a product through the manufacturing system, including control. CIM is an attempt to combine existing computer technologies in order to manage and control the entire business. CIM is the approach that many companies are using to get to the automated factory of the future.

As with the traditional manufacturing approaches, the purpose of CIM is to transform product designs and materials into salable goods at a minimum cost in the manufacture of that product. With CIM, the customary split between design and manufacturing functions is eliminated.

The element of CIM differs from the traditional job shop manufacturing system is the role the computer plays in the manufacturing process. Computer-integrated manufacturing systems are basically a network of computer systems tied together by a single integrated database. Using the information in the database, a CIM system can direct manufacturing activities, record results, and maintain accurate data. CIM is the computerization of design, manufacturing, distribution, and financial function into one coherent system.

The major element of a CIM system is a computer assisted design (CAD) system. CAD involves any type of design activity that makes use of the computer to develop, analyze, or modify an engineering design. The design-related tasks performed by a CAD system are:
- Geometric modeling;
- Engineering analysis;
- Design review and evaluation;
- Automated drafting.

Another major element of CIM is a computer-aided manufacturing (CAM) system. An important reason for using a CAM is that it provides a database for manufacturing the product. However, not all CAM databases are compatible with manufacturing software. The tasks performed by a CAM

system are:
- Numerical control (NC) or computer numerical control (CNC) programming;
- Computer-aided process planning (CAPP);
- Production planning and scheduling;
- Tool and fixture design.

New Words and Phrases

1. encompass [enˈkʌmpəs] *vt.* 包围，环绕，包含或包括某事物
2. coherent [kəuˈhiərənt] *adj.* 连贯的，一致的
3. geometric [ˌdʒiːəˈmetrik] *adj.* 几何学（的）
4. fixture [ˈfikstʃə] *n.* 夹具

Reading Material D

Virtual Reality

Virtual reality is a system that enables one or more users to move and react in a computer-simulated enviroment. Various types of devices allow users to sense and manipulate virtual objects as they would real objects. This natural style of interaction gives participants the feeling of being immersed in the simulated world. Virtual worlds are created by mathematical models and computer programs.

Virtual reality simulations differ from other computer simulations in that they require special interface devices that transmit the sights, sounds, and sensations of the simulated world to user. These devices also record and send the speech and movements of the participants to the simulation program.

In the future, your office may be less like a cubicle and more like the Holo-deck on "Star Trek". Computer scientists are already experimenting with the technology, called tele-immersion that will allow us to peer into the offices of colleagues hundreds of miles away and make us feel as if we are sharing the same physical space.

Computer scientists at several universities working on a project called the National Tele-Immersion Initiative have demonstrated a prototype system that enables a scientist working at his desk in Chapel Hill to see his distant colleagues on two screens mounted at right angles to his desk. It gives him the illusion of looking through windows into offices on the other side. And unlike videoconferencing, tele-immersion provides life-size 3-D images.

In the prototype system, each researcher is surrounded by a bank of digital cameras that monitor his movements from a variety of angles. The researchers also wear head-mounted tracking gear and polarized glasses similar to those used to view 3-D movies. When a researcher moves his own head, the view of his co-workers shifts accordingly. If he leans forward, for example, his colleagues appear closer, even though they are hundreds of miles away.

The scientists hope that tele-immersion will open up a host of other applications: for example, patients in areas could "visit" medical specialists in faraway cities.

New Words and Phrases

1. cubicle [ˈkjuːbikəl] n. 小卧室
2. Star Trek 《星际旅行》
3. holo-deck n. 全息驾驶舱
4. tele-immersion 远程投入，远程参与
5. videoconferencing [ˌvidiəuˈkɔnfərənsiŋ] n. 视频会议
6. co-workers n. 合作者
7. a host of 一大群
8. initiative [iˈniʃiətiv] n. 主动（性），第一步； adj. 自发的，初步的
9. a bank of 一系列；一排，一束

Part VI Document Retrieval

Unit 25 Introduction to Information Retrieval

In the mid of 1990s, along with the surge in electronic literature and the rapid development of the Internet, the focus of the literature retrieval began to tilt electronic database. In order to adapt to the development of the network times, in many fields "information retrieval" began to gradually replace "document retrieval".

Information retrieval refers to organizing and storing information in a certain way, and then finding relevant information according to information users' need, so its full name is "Information Storage and Retrieval", which is general information retrieval. The narrow sense of information retrieval only refers to the process after the second half part, that is, to find the information they need from information set, equivalent to information search people usually say.

Literature information retrieval is the guide of scientific research. In order to perform the valuable scientific research, we must depend on the literature search to the comprehensive gain literature information, and to understand each subject in the domain of the new problems and new ideas, so as to determine our own research points and the research targets. Scientific research first needs to master the material through the task survey. Literature retrieval will help us to understand the dynamic progress of subject research, to develop ideas, to avoid duplication, and to improve research level. The evaluation of scientific research achievements also needs to identify through the literature search, in order to make the correct conclusion. The ability of literature retrieval often affect the value of the scientific research achievements. Whether in topic sure stage, scientific experiment stage, or achievements summary stage, the literature retrieval is of great significance. It can say that, literature retrieval in scientific research work plays a very big role.

In determining the subject stage, we have to refer to a large number of literature material, in order to understand the history and present situation, the prospect and trend of the subject, and grasp what the predecessors have done, what others are doing, what problems exist in this topic, the experience and lessons, etc. On the basis of full survey, with the successful experience, failures and research methods, we will make the innovation of the conclusion different from the others, and work out specific scientific research plans. In scientific experiment stage, on the discovery process of the objective law, in order to solve the problems and difficulties, we must adopt the past experience of the reference to get the solution of difficult enlightenment. In results summary stage, in order to clarify the inheritance and creativity of the research, we must also widely collect a lot of paper, compare our results with the reported research achievements, and then make an objective evaluation to fully prove the accuracy and extension. Thus, in every stage of the scientific research, communication, accumulation, and inheritance with the use of literature information are always needed.

So, in scientific research process, literature retrieval not only can promote the rapid development and utilization of information resources, but also can help researchers inherit the results of the previous reference, to avoid repeated research and speed up the progress of the research. Specifically, the meaning and function of literature retrieval mainly include: (1) More specificly limit and determine the research subjects and assumptions; (2) Tell the researchers what research predecessors or others in this field have done; (3) May provide some valuable research ideas and methods of the study; (4) Present the appropriate modification suggestions, in order to prevent the difficulty of less than guesswork; (5) Make errors under control in the course of the study; (6) Provide background materials to explain the results.

Literature can be divided into zero times document, primary document, secondary document, and tertiary document. Zero times documents refer to the original documents without any process, such as the record of the experiment, manuscript, the original recording, the original video, notes of talks, etc. Primary document is the literature written by the author based on the study results. Most of the journal articles and published papers in science and technology conference belong to the primary documents. Secondary document refers to the product carried on the processing, refining, and compression of primary document. Tertiary document mainly includes the encyclopedia, dictionaries, etc.

Modern literature retrieval is mainly based on computer technology through the CD, online network, and other modern retrieval way. Information retrieval on the computer is a basic skill of modern scientific and technological personnel. It is extremely important to train the ability of scientific and technical personnel for future social and scientific research. A scientific researcher who is good at getting electronic information from the information system, will have more chances of success than people who don't have such ability. In the United States, interactive network retrieval expert has become one of 10 most popular careers.

Literature retrieval is a practical activity, which lets us be good at thinking, and through the regular practice, gradually grasp the law of literature retrieval, in order to quickly and accurately obtain literature. Generally speaking, the literature retrieval can be divided into the following steps: (1) Clear the purpose and requirements of searching; (2) Choose retrieval tools; (3) Determine the way and method of retrieval; (4) According to the literature clues, refer to the original documents.

At present, many research universities open the course named information retrieval and use. It is a course that tells us how to use electronic information resources under the environment of network. It mainly introduces the content of various kinds of electronic information resources on the Internet and retrieval technology, in order to let the students understand the various types of electronic information resources, understand and master the retrieval method of all kinds of database, electronic journals, electronic books, electronic newspapers, and other Internet information resources. Through the study of this course, students can master the skills of search and use of all kinds of electronic information resources to strengthen modern information consciousness and research ability, improve college students' ability to acquire literature and freely use all sorts of library resources and network information resources.

New Words and Phrases

1. equivalent to 相当于
2. reference ['refərəns] *n.* 参考，参照；参考书目；介绍信；*vt.* 引用
3. objective law 客观规律
4. stage [steidʒ] *n.* （进展的）阶段；时期
5. inseparable [in'sepərəbl] *adj.* 分不开的，不可分离的
6. original [ə'ridʒənl] *adj.* 最初的，本来的；原始的
7. gradually ['grædjuəli] *adv.* 逐步地，渐渐地
8. environment [in'vaiərənmənt] *n.* 环境，四周状况
9. database ['deitə,beis] *n.* 资料库，数据库
10. consciousness ['kɔnʃəsnis] *n.* 意识，意念

Notes

1. Information retrieval refers to organizing and storing information in a certain way, and then finding relevant information according to information users' need.

参考译文：文献检索是指将信息按一定的方式组织和存储起来，并根据信息用户的需要找出有关的信息过程。

2. On the basis of full survey, with the successful experience, failures and research methods, we will make the innovation of the conclusion different from the others, and work out specific scientific research plans.

参考译文：在充分调查的基础上，借鉴成功的经验、失败的教训和研究的方法，以便做出有别于他人的创新结论，制订出具体的科研计划。

3. Thus, in every stage of the scientific research reference, communication, accumulation, and inheritance with the use of literature information are always needed.

参考译文：由此可见，在科学研究的各个环节，自始至终都需要借鉴、交流、积累和继承，都离不开对文献信息的利用。

4. A scientific researcher who is good at getting electronic information from the information system, will have more chances of success than people who don't have such ability.

参考译文：一个善于从电子信息系统中获取信息的科研人员，必定比不具备这一能力的人有更多的成功机会。

5. Generally speaking, the literature retrieval can be divided into the following steps: (1) Clear the purpose and requirements of searching; (2) Choose retrieval tools; (3) Determine the way and method of retrieval; (4) According to the literature clues, refer to the original documents.

参考译文：一般来说，文献检索可分为以下步骤：（1）明确查找目的与要求；（2）选择检索工具；（3）确定检索途径和方法；（4）根据文献线索，查阅原始文献。

Exercises

1. Answer the following questions according to the text.

a) The narrow sense of information retrieval only refers to information search people usually say. Yes or no?

b) All aspects of scientific research, are inseparable from the use of literature information.

c) Literature can be divided into zero times document, primary document, and secondary document.

2. In the United States, interactive network retrieval expert has become one of 10 big popular careers.

a) yes   b) no   c) after ten years may be

3. Translate the following sentences into English.

a) 现代文献检索大多以计算机技术为手段,通过光盘和联机等现代检索方式进行信息检索。

b) 目前,很多研究型高校都开设了"信息检索与利用"课。

Unit 26   Professional Special Field Document Retrieval

Literature retrieval can be divided into manual retrieval and computer retrieval. Computer information retrieval has the advantages of high speed, less time-consuming, a wide range, and even getting access to the paper newly published in the foreign journals. At the same time, its content of retrieval is special. The disadvantages of computer information retrieval include certain limit of its backward time, relatively expensive cost of the retrieval, and certain restriction of retrieval time range. The computer information retrieval is mainly used to search digital literature information and dynamic information. The advantages and disadvantages of manual retrieval are just opposite to those of computer information retrieval. The advantage of manual retrieval is that the time and search range of retrieval are unlimited. But manual retrieval takes more time, with lower efficiency and less retrieval entrances. So the search results are often not as good as the computer information retrieval. Manual retrieval is mainly used in the paper printing literature, especially in early literature information search.

The professional document retrieval must be achieved by a more professional approach. There are various ways of retrieval, generally including:

1. According to the author to retrieve

Many retrieval systems contain author index, and the patent document retrieval systems contain the patentee index. Author, editor, translator, or the patentee all may be the author. Author can be a person, a unit, or an organization.

2. According to the title to retrieve

As the name suggests, the title refers to the name of a book, a publication, or an Article. The vast majority of retrieval systems provid an approach of retrieval according to the title name, such as the list of books, the list of directories, and so on.

3. According to the classification to retrieve

The meaning of classification is to distinguish and systematically arrange according to the nature of subjects, in order to enable users to find the literature through the system. Classification can embody the systematicness of the subject, reflect the membership, derivative and parallel relationship of discipline and things, to help us find the literature material from the subject scope and play the role of "analogy".

4. According to the theme to retrieve

The advantage of retrieval by the theme is that all literature material of a theme can be reflected, which facilitates readers to comprehensive research according to a certain question, a certain things and objects. We can get all literature of the same theme through the subject catalogue or index.

5. According to the citation to retrieve

The enclosed references or referenced literature, is one of the appearance characteristics of literature. The index system programmed with this citation is called citation index system. The researchers can retrieval relevant documents through the cited papers.

6. According to the serial number to retrieve

Some literature have specific serial numbers, such as the patent number, the report number, contract number, standard number, international standard book number, ISNN, etc. Serial number of documents is clear, short, and unique for distinguishing literature. All kinds of serial number index can provide a retrieval way according to the order of the literature information.

7. According to the code to retrieve

Retrieval by the code refers to the retrieval according to the specific code order with the use of a code of things into the index, such as molecular formula index, index of ring, and so on.

8. According to the special projects to retrieve

Some of the documents contain the specific information, such as terms, place names, organization names, person names, brand names, biological generic names, etc. Using this information to retrieve can solve some special problems.

China Journal Net is the most widely and frequently used network platform. A full understanding of China Journal Net can help us to improve the efficiency and quality of retrieval.

The full-text database for special topic of China Journal Net includes nine albums, such as polytechnic A, polytechnic B, polytechnic C, agriculture, medicine and health, wenshizhe, economy and politics and law, education and social science, electronic technology and information science. Retrieval methods include 13 search fields, such as title field, author field, keywords field, subject field, agency field, Chinese journal name field, Chinese abstract field, citation field, fund field, full text fields, year field, period number field, and ISSN field. Among them, the title, abstract, keywords, subject, and full text fields belong to the basic fields. There are three retrieval methods in China Journal Net full-text database. They are: 1) Classification navigation. Use the special column supplied by the system and classification navigation tree to search relevant literature, applicable to thoroughly investigate a subject of literature. 2) Primary retrieval. It is the use of a specified field to search for retrieval. 3) Higher retrieval. It sets up a multiple retrieval conditions and logical relationship, and determines the literatures which satisfy these conditions and logic relationship at the same time. The full-text database for special topic of China Journal Net has secondary retrieval function. It could carry on the second search again in the scope of recently retrieved results, so as to reduce the search range, and make the search results close to the requirements of the subject. Whether it is the primary or advanced search interface, as long as retrieval result number permitted, the secondary retrieval may operate repeatedly, until satisfying the research request. China Journal Net is provided with a full text browser function of full-text database, in order to facilitate browsing full texts of the database. The full text browser can be downloaded from the home page of the database.

New Words and Phrases

1. professional [prəˈfeʃənəl] *adj.* 职业性的，非业余性的
2. title [ˈtaitl] *n.* 标题，题目；书名；头衔；称号
3. patent [ˈpeitənt] *n.* 专利权；执照；专利品；*vt.* 授予专利；取得……的专利权
4. as the name suggests 顾名思义
5. advantage [ədˈvɑːntidʒ] *n.* 有利条件，优点，优势；利益，好处
6. frequent [ˈfriːkwənt] *adj.* 时常发生的，频繁的；惯常的；习以为常的
7. reduce [riˈdjuːs] *vt.* 减少；削减；缩小；使化为；使变为
8. requirement [riˈkwairmənt] *n.* 需要；必需品；要求；必要条件；规定
9. as long as = so long as 只要
10. facilitate [fəˈsiliteit] *vt.* 使容易；促进，帮助

Notes

1. Computer information retrieval has the advantages of high speed, less time-consuming, a wide range, and even getting access to the paper newly published in the foreign journals.

参考译文：计算机信息检索的优点在于速度快、耗时少、查阅范围广，甚至可以查到国外刚刚出版的期刊论文的信息。

2. But manual retrieval takes more time, with lower efficiency and less retrieval entrances. So the search results are often not as good as the computer information retrieval.

参考译文：但是手工检索耗时多、效率低、检索入口少，因此查找效果往往不如计算机信息检索的好。

3. The vast majority of retrieval systems provid an approach of retrieval according to the title name, such as the list of books, the list of directories, and so on.

参考译文：绝大部分检索系统中都提供按题名字顺检索的途径，如书名目录和刊名目录等。

4. Serial number of documents is clear, short, and unique for distinguishing literature.

参考译文：文献序号对于识别一定的文献，具有明确、简短、唯一性的特点。

5. A full understanding of China Journal Net can help us to improve the efficiency and quality

of retrieval.

参考译文：全面了解中国期刊网，对于提高检索效率和质量有重要意义。

6. The full-text database for special topic of China Journal Net has secondary retrieval function. It could carry on the second search again in the scope of recently retrieved results, so as to reduce the search range, and make the search results close to the requirements of the subject.

参考译文：中国期刊网专题全文数据库设有二次检索功能，可以在前次检索结果的范围内再次进行查找，以达到缩小检索范围，使检索结果逐步接近课题要求的目的。

Exercises

1. Answer the following questions according to the text.
a) Author can be a person, and also can be a unit or organization. Yes or no?
b) China Journal Net is the most widely and frequently used network platform. Yes or no?
c) The full-text database of China Journal Net includes ten albums. Yes or no?
d) Retrieval methods include 13 search fields. Yes or no?
2. In the retrieval method of China Journal Net, which fields belong to the basic fields?
3. Translate the following sentences into English.
a) 我们通过主题目录或索引，即可查到同一主题的各方面文献资料。
b) 不管是初级还是高级检索界面，只要检索结果的篇数允许，二次检索可以反复执行，直到满足研究要求为止。
4. 在中国知网 CNKI 上查找本校某教授 2010～2012 年在期刊上发表的文章有多少篇，其中核心期刊多少篇。并列举出该教授以第一作者发表的核心期刊的篇名、刊名及年/期。
5. Please use your two frequently used search engines to find professional information, write clear process and result, and make the appraisal of the retrieved results.

Appendix

Appendix A 课文译文及练习答案

第 I 部分 机械制造基础

第 1 单元 工程材料的分类

材料有多种分类方法。科学家常把材料按其状态分成：固态、液态和气态材料，也可分成有机材料和无机材料。按工业用途，材料可分成工程材料和非工程材料。工程材料是用于制造和做成产品零件的材料，非工程材料则是指化学制品、燃料、润滑剂和那些参与制造过程但并不被做成产品零件的材料。

我们将工程材料分成四类：金属、陶瓷、高聚物和复合材料．

1. 金属

金属一般被定义为其氢氧化物为碱性的物质（如钠、钾），金属可以是由单一元素构成的纯金属，也可以是由两种或两种以上金属元素组成的化合物，这种化合物称为合金。

"合金"这个术语用于识别任何金属系统。在冶金学中，合金是由两种或两种以上的元素均匀混合、具有金属特性的物质。在所构成的元素中，必须有一种是金属。例如，普通碳钢主要是由碳和铁组成的合金，当然还包括其他的一些元素。然而，出于商业目的，普通碳钢没有被归类于合金钢。

钢、铝、镁、锌、铸铁、钛、铜、镍等金属和合金的共性有：导电性和导热性好，强度和硬度较高，有一定的韧性、可成型性和抗冲击性。它们常用于结构件和承载件。尽管纯金属也偶尔使用，但合金更常用于设计满足特殊性能的需求或者达到更好的性能组合。

2. 陶瓷

砖、玻璃、餐具、绝缘体、磨料等陶瓷类材料的导电性和导热性差，虽然具有高强度、高硬度，但韧性、可成型性和抗冲击性差。因此，与金属相比，陶瓷很少用于结构件或承载件。然而，许多陶瓷具有极好的耐高温性和一定的耐腐蚀性以及许多非同寻常且令人满意的光学、电学和热学性能。

3. 聚合物

聚合物包含橡胶、塑料和各种粘胶剂。它们是由来自石油或农产品的有机分子通过聚合形成的巨大分子结构所产生的。聚合物的导电性和导热性差，强度低，不适合高温环境。有一些聚合物有极好的韧性、可成型性和抗冲击性，而另一些聚合物性能则相反。聚合物通常较轻且抗腐蚀性极好。

4. 复合材料

复合材料由两种或两种以上的材料组成，其性能绝非任何一种单一材料所能拥有。如混凝土、夹板和玻璃纤维，尽管未经加工，却是典型的复合材料。采用复合材料，人们能生产出集质轻、坚固、易延展、耐热性强等性能于一体的材料，这些材料用其他手段是不能生产的；也可以制造出既有高硬度又有抗冲击性的切割工具，否则仅有高硬度（加工时）则容易碎裂。

练习答案

2. a) F   b) F   c) T T F   d) T F

第 2 单元　极限尺寸、配合与公差

基本术语

配合：两个相配零件尺寸之间的差别所产生的作用。

基本尺寸：作为基准来确定极限尺寸的尺寸，也称为名义尺寸。

实际尺寸：通过测量得到的零件尺寸。

极限尺寸：包含实际尺寸在内的两个极限的允许尺寸，两个极限尺寸称为最大极限尺寸和最小极限尺寸。

公差：最大极限尺寸和最小极限尺寸之差（也等于上偏差与下偏差之间的代数差）。

上偏差：最大极限尺寸和对应基本尺寸的代数差。

下偏差：最小极限尺寸和对应基本尺寸的代数差。

配合种类

按孔或轴的实际尺寸，配合可划分如下：

(1) 间隙配合：在相配零件间总是存在间隙的配合，此时，孔的公差带完全位于轴的公差带之上。

(2) 过盈配合：在相配零件间总是存在过盈的配合，此时，孔的公差带完全位于轴的公差带之下。

(3) 过渡配合：取决于精加工产品的实际尺寸，在相配零件间存在间隙或存在过盈的配合。

应用于配合系统的不同方法中，主要有基轴制和基孔制。在基轴制中，将不同的孔与单一基本尺寸的轴相配，可得到不同的间隙值和过盈量，轴的上偏差是 0（符号 h）。在基孔制中，将不同的轴与单一基本尺寸的孔相配，可得到不同的间隙值和过盈量，孔的下偏差是 0（符号 H）。

通常，制造规定公差的轴比制造同样公差的孔容易。因此，在现代工程设计中，基孔制系统应用最广泛，我们主要讨论这种系统。但设计者应决定采用何种系统以确保完全互换性。

公差和配合的符号

公差用一个字母（有时两个字母）、一个符号和一个数字表示。孔用大写字母，轴用小写字母，字母符号表明了公差带相对于基本尺寸的零线的位置。数字代表了公差带的值，称为公差等级。公差等级和位置是基本尺寸的函数。因此，公差尺寸用基本尺寸加上字母和数字表示，如 $\phi50H7$ 和 $\phi50g6$。

公差系统的基本内容

对所有的工业测量,标准参考温度是 293K。基本尺寸从 1mm 到 500mm 再划分为 13 个等级。从 500mm 到 3150mm,有 8 个名义等级。对每个名义等级,有 20 级公差,标记为 IT01、IT0 和 IT1、…、IT17、IT18,这称为标准公差。(IT 代表 ISO 公差等级系列,ISO 代表国际标准化组织)。

在工程制图中,名义尺寸必须带有公差:
(a) 加工零件的功能或经济性要求一定的极限尺寸;
(b) 零件要求有配合;
(c) 零件单独精加工并无需后续加工就直接装配;
(d) 零件要求能互换,如:备件;
(e) 零件必须要有公差,可以被夹在夹具上进行精加工。

其余情况下,自由公差就能满足要求。

公差选择时,应使工件满足相关应用场合的需要,并能保证互换性的要求。公差越小,生产成本越高。

公差等级的选择

IT01 到 IT7 主要用于量规,其中 IT01 到 IT4 要通过研磨、珩磨和精磨。

IT5 到 IT11 主要用于经过切削加工的工件的配合,如精车、铣削、成形加工、刨削、磨削和铰削。

IT12 到 IT18 级适用于不进行切削加工表面较粗的场合,如锻造、滚压、铸造、压力加工和拉丝。

练习答案

1. a) 名义尺寸　　b) 实际尺寸和极限尺寸　　c) 上偏差和下偏差
   d) 间隙配合和过盈配合　　　　　　　　　 e) 国际标准化组织

第 3 单元　金属的热处理

将固态的金属加热、冷却以改变其物理性能的过程称为热处理。经热处理后,钢可以硬化以抵抗切削和磨损,也可以软化以便加工。使用适当的热处理,可以除去内部应力,细化晶粒,提高韧性,使零件表面硬、内部韧。

下面主要讨论普通碳素钢的热处理。在这个过程中,冷却速度是控制因素,在临界范围以上的快速冷却将导致组织硬化,而非常慢的冷却,其效果则相反。

淬火:在任何一种热处理的操作中,加热的速度都是非常重要的。热量从钢的外部以一个确定的最大速度向内部传递,如果加热的速度太快,零件的外部温度就会高于其内部温度,这样将会很难获得内外均匀一致的组织结构。

通过热处理获得的硬度主要与以下三方面因素有关:

1. 淬火速度;
2. 碳的含量;
3. 工件的尺寸。

普通低碳钢和中碳钢为了淬硬,应该采用快速淬火工艺,通常采用水作为冷却介质。对高碳钢和合金钢，则采用油冷，油冷的淬硬作用比不上水冷。如果要求严格冷却，就必须采用盐水。

直接冷却后的钢所能获得的最大硬度在很大程度上由碳的含量决定，因此，低碳钢即使在热处理后也不可能达到太高的硬度。碳钢一般也被称为浅硬化钢。不同钢的淬火温度不一样，这由碳的含量所决定。

通常使钢淬火变硬的温度，称为淬火温度。它通常要高于加热转变临界温度 10~38℃，在这个温度时金属的组织结构就会发生变化。

回火：淬火使高碳钢和工具钢变得极其硬而且脆，大多数情况下不能直接使用。通过回火，淬火过程中产生的内应力得以消除。回火提高了淬火零件的韧性，也使材料具有更大的塑性或延展性。

退火：退火是将金属稍微加热到临界温度以上后，很缓慢地冷却。退火处理能够减轻金属内部由于先前热处理、切削加工或其他冷加工所造成的内应力和应变。钢的种类决定着退火加热的温度，加热的温度也与退火的目的有关。

工业中应用的退火主要有三种：（1）完全退火，（2）低温退火，（3）球化退火。

完全退火用来最大限度地降低钢的硬度，以改善它的切削加工性能，消除内应力。低温退火也称去应力退火，它的目的主要是消除在冷加工和机械加工过程中产生的内应力。球化退火是使钢中生成一种特殊的晶粒结构，这种结构相时较软而易于加工。这种工艺一般用于改善高碳钢的切削加工性能和用于拉丝工艺的热处理。

正火：正火操作是用来消除金属由于热加工、冷加工及机械加工过程中产生的内应力的过程。正火是把钢加热到临界温度以上 30~50℃，保温一段时间后空冷。正火通常应用于低、中碳钢和合金钢。正火可以消除先前热处理所留下的各种影响。

练习答案

1. a）Annealing/three　b）toughness/plastic/ductile
2. a）热处理　b）淬火速度　c）低温退火　d）球化退火　e）去应力退火

第 4 单元　成　　形

成形可以定义为一种通过材料的塑性变形获得所需尺寸和形状的工艺过程。在此工艺中引起的应力大于材料的屈服强度，但小于材料的断裂强度。加载的类型可以是拉应力、压应力、弯曲应力、切应力，或者是这些应力的组合。这是个很经济的方法，因为可以获得所需的形状、尺寸和表面粗糙度而无需使材料有任何大的损失。此外，一部分输入的能量在通过应变硬化提高产品强度时得到了卓有成效的利用。

成形工艺可以分为以下两大类，即冷成形和热成形。如果热加工温度高于材料的再结晶温度，那么这一过程就被称为热成形，否则被称为冷成形。材料流动应力的作用在再结晶温度之上或之下全然不同。在热加工过程中，可以产生大的塑性变形而无明显的冷作硬化。这一点很重要，因为大量的冷作硬化会使材料变脆。两种成形方法的摩擦特性也完全不同。例如，冷成形的摩擦系数一般为 0.1 左右，而热成形的摩擦系数可以高达 0.6。此外，热成形降低了材料的强度，以至于使用具有适度功率的机器就可以加工很大尺寸的产品。

典型的成形方法有轧制、锻造、拉延、深拉、弯曲和挤压。为了更好地理解各种成形操作的机械学原理，我们将简要讨论每一种方法。

轧制
在这一工艺中，通过一个位于两个动力驱动轧辊之间的调整过的轧辊缝，利用摩擦来拉伸工件。产品的形状和尺寸由轧辊及其轮廓之间的间隙来决定。这是一种很有用途的工艺，用于生产金属薄板和各种常用断面，如铁轨、槽钢、角钢和圆钢。

锻造
锻造时，材料在两个或多个模具间受到挤压以改变其形状和尺寸。根据情况不同，模具可以是开式或闭式。

拉延
在这一工艺中，由于工件被拉过模具的锥形孔，从而导致金属丝、条钢或钢管的截面减小。当截面需要减小很多时，也许有必要通过几个阶段来完成此操作。

深拉
在深拉中，杯形产品是在一个凸模和一个凹模的帮助下由一决金属薄板获得的。坯料被压板夹住以避免产品出现缺陷（折皱）。

弯曲
顾名思义，这是一道塑性弯曲金属薄板以获得所需形状的工艺。这套工艺由一套设计适当的凸模和凹模来完成。

挤压
这是一道基本上类似于闭式模具锻造的工艺。但是在工艺中，工件被压进一个封闭空间，迫使材料从模具的适当开口处挤出。在这一工艺中，只可以制造出具有固定截面（模具出口截面）形状的工件。

热、冷成形的优点和缺点
既然我们已经谈及了各类金属加工工艺，现在我们应该给热加工和冷加工工艺一个总体评价了。这一讨论将有助于为给定的情况选择适合的加工条件。

由于活跃晶粒在加工温度范围内会长大，故在热加工中就可以适当地控制晶粒的大小。结果，由于没有冷作硬化，因而也不需要进行昂贵耗时的中间退火。当然，在一些操作（如拉延）中冷作硬化还是需要的，能用以提高强度；在这些情况下，热加工几乎没有优势。除此之外，要成功地完成某些工序，冷作硬化可能是必不可少的（如在深拉中，冷作硬化可防止应力最大的底部过渡圆角附近的材料断裂）。在热加工条件下，可以加工大件产品和高强度材料。因为温度的上升可使材料的强度下降，进而降低了工作载荷。此外，对于大多数材料来说，延展性随温度的上升而增大，因此易碎的材料也可采用热加工工艺。但是应该记住某些材抖（如含硫磺的钢）在提高温度时会变得更脆。当需要非常精确地控制尺寸时，热加工就不太适合，因为金属的表面会生成氧化皮而收缩或损失。而且，由于氧化物的形成及其皮层剥落，零件的表面粗糙度也不好。

冷加工的主要优点是经济、操作更为迅速、容易，因为无须安排额外的加热和处理。另外，在加工过程中由于冷作硬化的作用，材料的机械性能一般也得到改善。再者，控制了晶格排列的方向即可增加材料的强度特性。然而，冷加工除了其他一些限制条件外（如难以加工高强度和脆性材料以及大尺寸的产品），该工艺还有一个缺点，即无法防止材料抗腐性能的明显下降。

练习答案

1. 1-c 2-b 3-a 4-f 5-e 6-d 7-h 8-g 9-j 10-i

阅读材料 A　钢的种类

通常钢分为两大类：碳素钢与合金钢。碳素钢仅含有铁和碳，而合金钢则含有某些其他"合金元素"，如镍、铬、锰、钼、钨、钒等。

1. 碳素钢

（1）低碳钢（含碳0.05%～0.15%），这种钢还称为"结构钢"。

（2）中碳钢（含碳0.15%～0.60%）。

（3）高碳钢（含碳0.60%～1.50%），这种钢也称为"工具钢"。

2. 合金钢

（1）特种合金钢，如镍钢、铬钢。

（2）高速钢，还称为自硬钢。

碳钢的性能仅取决于钢中的含碳量。低碳钢质软，故用于制造没有强度要求的螺栓和机器零件。

中碳钢是较高级钢，强度比低碳钢高，切削加工也较低碳钢困难。

高碳钢可借助加热到一定温度，然后通过在水中快速冷却的方法使之硬化。钢的含碳量越高，冷却越快，钢就变得越硬。这种钢，因为其强度、硬度高，可用于生产刀具和机器工作零件。但对于某些特殊用途的零件来说，比如齿轮、轴承、弹簧、轴以及金属丝等，因为碳钢不具备这些零件所需要的性能，所以常常不予采用。

对这类零件可采用某些特殊的合金钢，这是因为合金元素能提高钢的韧性、强度和硬度。有些合金元素能使钢耐腐蚀，所以这种钢被称为不锈钢。

耐热钢是通过加入钨和钼而制得，而锰能增加钢的耐磨性能。钒钢能够抗腐蚀，并能承受冲击和振动。

用含有钨、铬、钒、碳的高速钢制作的刀具能够比碳素工具钢以高得多的速度进行切削加工。

阅读材料 B　尺寸和公差

在图样上标注尺寸时，尺寸线上标注的数字仅仅表示近似的尺寸，而不表示任何精度，除非设计人员加以说明。这个数字称为公差尺寸。零件的公差尺寸是设计人员根据设计工艺的需要而制定的一个恰当的尺寸值。然而，用已知的任一种制造工艺都几乎不可能将零件加工到百分之百精确的尺寸。即使用手工的方法感觉将一个零件加工到精确的尺寸，但用高精度的测量设备进行测量也会发现这种感觉是不正确的。因此，在工程实际中常允许在公差尺寸左右有一个变化范围，这称为公差。尺寸的公差也能说明精确的程度。例如一根轴的公差尺寸63.5mm，如果允许的变化范围是±0.08mm，该尺寸就可以标注为63.5mm±0.08mm。

在工程中，当所设计的某一产品是由许多零件所组成时，这些零件以某种形式彼此配合。在装配时，考虑到两零件的配合类型是很重要的，因为这将影响到零件与零件之间的运动形式。以轴和孔配合为例，最简单的情况是轴的尺寸比孔的尺寸小，轴与孔之间会存在间隙，这样的配合称为间隙配合。反过来，若轴的尺寸比孔的尺寸大，则称为过盈配合。

紧公差的尺寸表示该零件必须与其他零件恰当配合。两者所标公差必须要与设计允差、制造工艺的能力以及获取最大利润的最小生产与装配成本一致。一般来说，零件的成本是随公差的减小而上升的。如果一个零件要加工几个或更多的表面，允许的偏差与公差尺寸的差值很小时，那么成本将是非常大的。

容差有时与公差相混淆，两者的意义是完全不同的。容差是在两配合零件间允许的最小间隙，表示配合允许的最紧的状态。假设直径为 $1.498^{-0.000}_{-0.003}$ 的轴，要与直径为 $1.500^{+0.003}_{-0.000}$ 的孔才相配合。孔的最小尺寸为 1.500，轴的最大尺寸为 1.498。因此容差为 0.002，根据轴的最小尺寸和孔的最大尺寸可得最大间隙为 0.008。

公差可以是单向的或双向的。单向公差只允许在公差尺寸或基准尺寸的一个方向上有变化量。参考前面的例子，孔的尺寸为 $1.500^{+0.003}_{-0.000}$，这就是一个单向公差。在这里公差尺寸 1.500 可在 1.503 到 1.500 之间变化。如果给定尺寸为 1.500 ± 0.003，这就是双向公差。也就是说，它可在公差尺寸的两边变化。双边公差中，极限的变化范围可以是一致的，如 30.00 ± 0.02，尺寸变化从 30.02 到 29.98。另外偏差也可以不同，如 $30.00^{+0.05}_{-0.10}$，尺寸变化从 30.05 到 29.90。有时，公差尺寸可在极限之外。例如，一个尺寸的变化范围为 29.95 到 29.85，可以写成 $29.95^{+0}_{-0.10}$ 或 $30.00^{-0.05}_{-0.15}$，因为后一种形式包含公差尺寸 30，所以首选后一种形式。单向公差系统允许容差与配合类型不变化的情况下改变公差。而对双向公差系统来说，如果不改变两个配合零件中的一个或两个的公差尺寸，则改变公差是不可能的。在大规模生产中要求互相配合的零件是可互换的，因此单向公差更适合。两配合零件要想做到静配合或压配合，公差必须是零公差或负公差。

阅读材料 C 金属热加工

我们都知道，铸造就是将熔化的金属浇注进预定铸模中，并使其冷却成形的一个机械加工过程。当金属没有按预定形状进行铸造的时候，就要通过机械加工方法使其成形。当决定要采用哪种方法的时候，应该综合考虑多种因素。如果零件的形状非常复杂，就必须采用铸造的方法，这样可以避免昂贵的机加工费用。另一方面，在指定的零件中，如果材料的强度和质量是主要的因素，这时采用铸造就不合适。正是由于这个理由，在飞机制造业上很少使用铸钢。

金属加工有以下三种基本的方法。它们是热加工、冷加工和挤压。虽然在加工某一个工件的时候，既可以用热加工也可以用冷加工，但还是应该根据所采用的金属和对零件的要求来选择加工的方法。

几乎所有的钢都是由工业纯铁经过热加工得到的，把钢再进行热加工或冷加工就可以制成最终产品。当把铁锭从模具中取出来时，它的表面是凝固的，但其内部却仍是液态。然后再将铁锭放进均热炉内，减缓热量的损失，使内部熔化的金属渐渐地凝固下来。此过程结束之后，纯铁各处的温度都相等，然后通过热轧使它的尺寸变得适中，这样就可以更加方便地处理。

热加工可以使工业纯铁被加工成理想的形状。它工作的温度通常是比较高的。在高温下，钢会出现热膨胀以及其表面被氧化的现象，我们并不希望以此作为终加工。所以，大多数含铁金属要在热加工之后安排冷加工，目的是为了提高金属的表面光洁度。

热加工的主要原理是使材料发生塑性变形。它所需要的力一般要比冷加工小。在热加工的过程中并没有改变金属的机械特性，原因在于发生变形的温度高于金属的再结晶温度。金属在高于再结晶温度下发生的塑性变形是没有任何应变强化的。事实上，通过加热通常会使金属的屈服强度降低，因此，对金属进行热加工而不使其断裂是有可能的。

阅读材料 D　表面粗糙度

在零件的实际加工中，想要得到理想的表面质量是不可能的，因此在机械设计的图样中，一定要标注出零件允许的表面条件信息。表面条件是所采用加工方式的函数。图 ID-1 形象地表示了影响表面粗糙度的各相关参数。

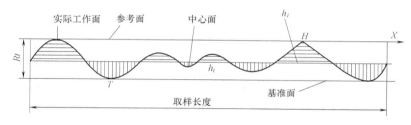

图 ID-1　表面粗糙度

实际工作面：实际工作表面。
参考面：指工件不规则的表面区域，它是由实际工作面的最高点 H 来确定的。
基准面：平行于参考面且通过实际工作面的最低点 T。
中心面或平均面：在取样长度内，中心面把表面波峰与波谷分成上、下两部分。使中心面以上部分的实体材料之和等于中心面以下部分的实体材料之和。
峰谷高度 Rt：表示基准面和参考面之间的距离。
平均粗糙度指数 Ra：实际工作面上各点至中心面的距离的绝对值之和的算术平均值。其表达式为

$$Ra = \frac{1}{L}\int_{x=0}^{x=L} |h_i| \mathrm{d}x$$

在工程图样中表面粗糙度有好几种表达方式。IS：696—1972 对其做了规定。表面粗糙度的基本符号如图 ID-2a 所示，它由两条长度不等的线表示。基本符号上加一短划线，表示表面是用去除材料的方法获得的，如图 ID-2b 所示。基本符号上加一小圆，说明表面是用不去除材料的方法获得的，如图 ID-2c 所示。当某些特殊表面有特别的要求时，在基本符号的稍长的一边加一横线即可，如图 ID-2d 所示。表

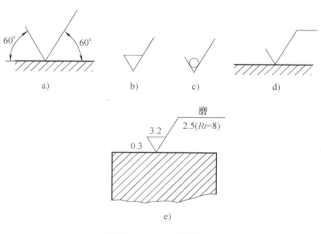

图 ID-2　表面粗糙度

面粗糙度和相关数据的标注位置，如图 ID-2e 所示。

第Ⅱ部分 机　　床

第 5 单元 车　　床

用于车外圆、端面和镗孔等加工的机床称为车床。车削很少在其他种类的机床上进行，因为其他机床都不能像车床那样方便地进行车削加工。因为车床除了用于车外圆外还能用于镗孔、车端面、钻孔和铰孔，所以车床的多功能性可以使工件在一次定位安装中完成多种加工。

图 5-1 中标出了车床的主要部件：床身、主轴箱组件、尾座组件、拖板组件、变速齿轮箱、丝杠和光杠。

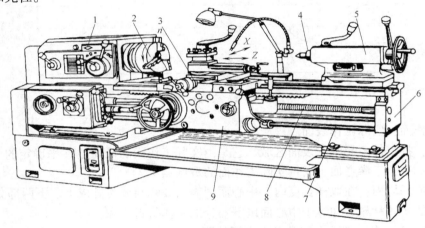

图 5-1　普通车床结构图
1—主轴箱　2—卡盘　3—刀架　4—顶尖　5—尾座
6—床身　7—光杠　8—丝杠　9—溜板箱

床身是车床的基础件。它通常由经过充分正火或时效处理的灰铸铁或球墨铸铁制成。它是一个坚固的刚性框架，其他所有主要部件都安装在床身上。通常在床身上面有内外两组平行的导轨。因为其他部件安装在导轨上并（或）在导轨上移动，所以导轨要经过精密加工，以保证其装配精度。同样，在操作中应该小心，以避免损伤导轨。导轨上的任何误差，通常会使整个机床的精度遭到破坏。大多数现代车床的导轨要进行表面淬火处理，以减小磨损和擦伤，并具有更大的耐磨性。

主轴箱安装在床身一端内导轨的固定位置上。它提供动力，使工件在各种速度下旋转。它主要由一个安装在精密轴承中的空心主轴和一系列与卡车变速箱类似的变速齿轮所组成，通过变速齿轮，主轴可以在许多种转速下旋转。

因为车床的精度在很大程度上取决于主轴，所以主轴的结构尺寸较大，通常安装在紧密配合的重型圆锥滚子轴承或球轴承中。

主轴的内端从主轴箱中凸出，其上可以安装多种卡盘、花盘和挡块。而小型的车床常带有螺纹截面供安装卡盘之用。很多大车床使用偏心夹或键动圆锥轴头。

尾座组件主要由三部分组成。底座安装在床身的内侧导轨上，并可以在导轨上作纵向移动,底座上有一个可以使整个尾座组件夹紧在任意位置上的装置。尾座安装在底座上,可以沿键槽在底座上横向移动,使尾座与主轴箱中的主轴对齐并为切削圆锥体提供方便。尾座组件的第三部分是尾座套筒,它是一个直径约为 2~3in 的钢制空心圆柱轴。通过手轮和螺杆,尾座套筒可以在尾座体中纵向移入和移出几英寸。

拖板组件用于安装和移动切削刀具。拖板是一个相对平滑的 H 形铸件,安装在床身外侧导轨上,并可在上面移动。

大多数车床的刀架安装在复式刀座上。刀座的底座安装在横拖板上,可绕垂直轴和上刀架转动。上刀架安装在底座上,可用手轮和刻度盘控制一个短丝杠使其前后移动。

溜板箱装在大拖板前面,通过溜板箱内的机械装置可以手动和动力驱动大拖板以及动力驱动横拖板。通过转动溜板箱前的手轮,可以手动操作拖板沿床身移动。手轮的另一端与溜板箱背面的小齿轮连接,小齿轮与齿条啮合,齿条倒装在床身前上边缘的下面。

利用光杠可以将动力传递给大拖板和横拖板。光杠上有一个几乎贯穿于整个光杠的键槽,光杠通过两个转向相反并用键联接的锥齿轮传递动力。

现代车床有一个变速齿轮箱,齿轮箱的输入端由车床主轴通过合适的齿轮传动来驱动。齿轮箱的输出端与光杠和丝杠连接。主轴就是这样通过齿轮传动系驱动变速齿轮箱,再带动丝杠和光杠,然后带动拖板,刀具就可以按主轴的转数纵向地或横向地精确移动。

练习答案

1. a) 车外圆、车端面和镗孔 b) 主轴箱组件 c) 尾座组件 d) 拖板组件 e) 丝杠和光杠

第 6 单元 磨削分类与无心磨削

一般来说,磨削被用来作为精加工工序,使产品达到所要求的表面光洁度、正确的尺寸和精确的外形。但是,近期的研究表明磨削也可以像车削、铣削等加工方法那样很经济地用于大量去除不需要的材料。

磨削加工的分类

磨削加工可以根据被磨制工件的表面形状、磨床的类型或磨削的产品类型进行分类。根据表面的类型和磨床的类型分类如下:

1. 平面磨削:用于磨削平面。
2. 内圆、外圆磨削:用于磨削内、外圆柱表面。
3. 无心磨削:用于磨削内、外圆柱表面。在这一加工过程中,使用的磨床不同于常规的内、外圆磨床。
4. 成形磨削:包括齿轮磨削、螺纹磨削和花键轴磨削等。
5. 磨削切割加工:使用高速旋转的薄型砂轮加工金属和非金属材料。这种加工方法取决于磨粒的切割作用,切割过程中产生的热量有助于切割。
6. 砂带磨削:这种方法在磨削加工中被看做是一个重要的加工方法。由于砂带易于贴合零件的形状,因此它可以用来磨削平滑的、圆柱的和曲面的形状。
7. 手工磨削加工:在这一加工过程中,手持工件或砂轮移动并加工。所用的机械有台式磨床、便携式磨床、模具磨床等。

然而，针对很多零件已设计出专用磨床，如曲轴磨床、凸轮轴磨床等。

在各种磨削加工中，可以根据磨制产品的类型或磨床控制的类型进行选择。例如：在圆柱磨削加工中，可以使用下列各种不同类型的磨床。

1. 普通外圆磨床。
2. 重型平面或轧辊磨床。
3. 万能外圆磨床。
4. 计算机数控外圆磨床。
5. 仿形磨床，如计算机控制的凸轮磨床。
6. 曲轴磨床。

在现代计算机数控机床中，已经实现了砂轮自动修整和工件尺寸自动控制。

无心磨削

在无心磨削加工中，磨削圆柱时工件既不像外圆磨床那样用顶尖顶住圆柱的两个中心孔，也不使用卡盘，圆柱形工件被支撑在砂轮、导轮和工件支架上，工件的旋转速度由导轮的表面速度控制。通过调节工件支架，工件的中心就可以保持在砂轮和导轮的中心连接线的上方。工件支架表面相对于中心连线有一倾角，其斜倾角度和工件中心的高度是获得精确的圆柱表面的重要参数。

外圆无心磨削

为了加工不同外形的产品，工业上使用四种外圆无心磨削。

1. 纵向进给无心磨削。
2. 横向进给无心磨削。
3. 纵向定程进给无心磨削。
4. 横向进给和纵向定程组合无心磨削。

纵向进给加工用于普通圆柱形工件。导轮轴相对于砂轮轴稍有倾斜。这样工件就可获得两种类型的运动，即（1）绕自身轴的旋转，（2）平行于砂轮轴的直线运动。

横向进给加工用于阶梯圆柱体工件加工，这种工件不能贯穿进给。在这种情况下，导轮回退，送进工件，然后推进导轮进行磨削加工。

内圆无心磨削

内圆磨削有两种加工方法。在第一种加工方法中，管状工件放置在导轮、支撑轮和加压辊之间，导轮、工件和砂轮的中心全都在同一条直线上。这种磨削就称为同心内圆无心磨削。

在第二种加工方法中，工件也是支撑在导轮、支撑轮和加压辊之间的，但是砂轮的中心并不位于导轮中心和工件中心的连接线上，这种磨削称为偏心内圆无心磨削。

在第一种加工方法中，即使是非常薄的管状工件，其壁厚也能精确地研磨。

练习答案

1. a) 横向进给无心磨削　　　b) 螺纹磨削　　　c) 纵向进给无心磨削
   d) 研磨轮的轴　　　　　　e) 纵向定程进给无心磨削　　f) 万能外圆磨床
   g) 计算机数控外圆磨床　　　h) 曲轴磨床

第 7 单元 铣削

铣削是机械加工的一种基础方法。在这一加工过程中,当工件沿垂直于旋转刀具轴线方向进给时,在工件上形成并去除切屑从而逐渐地铣出表面。有时候,工件是固定的,而刀具处于进给状态。在大多数情况下使用多齿刀具,金属切削量大,只需一次铣削就可以获得所期望的表面。

在铣削加工中,使用的刀具称为铣刀。它通常是一个绕其轴线旋转并且周边带有同间距齿的圆柱体,铣刀齿间歇性接触并切削工件。在某些情况下,铣刀上的刀齿会高出圆柱体的一端或两端。

因为铣削切削金属速度很快,并且能产生良好的表面光洁度,故特别适合大规模生产加工。为了实现这一目的,已经制造出了质量一流的铣床。然而在机修车间和工具模具加工中也已经广泛地使用了非常精确的多功能通用铣床。车间里拥有一台铣床和一台普通车床就能加工出具有适合尺寸的各种产品。

铣削操作类型

铣削操作可以分成两大种类,每一类又有多种类型。

1. 圆周铣削 在圆周铣削中,使用的铣刀刀齿固定在刀体的圆周面上,工件铣削表面与旋转刀具轴线平行,从而加工表面。使用这种方法可以加工出平面和成形表面,加工中表面横截面与刀具的轴向外轮廓一致。这种加工过程常被称为平面铣削。

2. 端面铣削 铣削平面与刀具的轴线垂直,被加工平面是位于刀具周边和端面的齿综合作用形成的。刀具周边齿完成铣削的主要任务,而端面齿用于精铣。

圆周铣削和端面铣削的基本概念如图 7-1 所示。圆周铣削通常使用卧式铣床,而端面铣削则既可在卧式铣床又可在立式铣床上进行。

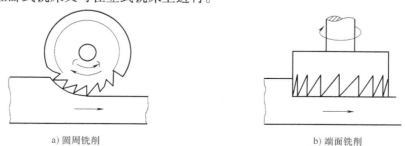

a) 圆周铣削   b) 端面铣削

图 7-1 圆周铣削和端面铣削中刀具和工件的关系

铣刀

铣刀分类有多种方法,一种方法是根据刀具后角将铣刀分为两大类。

1. 仿形铣刀 每个刀齿在切削刃的背面磨了一个很小的棱面形成后角,切削刃可以是直线或曲线。

2. 成形或凸轮形后角铣刀 每个齿的横截面在切削刃的背面呈偏心曲线状,以产生后角。偏心后角的各面与切削刃平行,具有切削刃的相同形状。这种类型的铣刀仅需磨削齿的前刀面就可以变得锋利,而切削刃的外形保持不变。

铣刀的另一种分类方法是根据铣刀安装的方法进行分类。心轴铣刀带有一个中心孔以使铣刀安装在心轴上。带柄铣刀有一锥柄或直柄轴，含锥形轴柄的铣刀可以直接安装在铣床的主轴上，而直柄轴的铣刀则是夹持在卡盘里。平面铣刀通常用螺栓固定在刀轴的末端上。

铣刀的类型

圆柱形铣刀是在圆周上有直的或螺旋形的齿的圆柱形或盘形铣刀。它们可以用来铣削平面，这种铣削称为平面铣削。螺旋形的铣刀上的每个齿是逐渐地接触工件的，在给定的时间内，一般有多齿进行铣削，这样可以减少振动，获得一个较平滑的表面。因此，与直齿铣刀相比，这种类型的铣刀，通常使用得更多。

侧刃铣刀的齿除了在圆柱刀体的一端或两端向径向延伸之外，与圆柱形铣刀是相似的。侧刃铣刀的刀齿既可以是直线的，也可以是螺旋形的，这种铣刀一般较窄小，具有盘形的形状。在跨式铣削加工中，常常将两个或更多的侧刃铣刀同时相间地安装在一个刀杆上同步并行切削。

双联槽铣刀是由两个侧刃铣刀组成的，但是在铣槽时，作为一组铣刀进行操作。在两个铣刀之间添加一些薄垫片，以调整之间的间距。

错齿铣刀是较薄的圆柱形铣刀，刀上有相互交错的刀齿，相邻刀齿具有相反的螺旋角。这种铣刀经研磨后仅用于周铣，在每个齿突出的一边，留有供切屑排出的缝隙。这种类型的铣刀可用于高速切削，在铣削深槽时可以发挥独特的作用。

开槽铣刀是一种薄型的圆柱形铣刀，厚度一般为 1/32～3/16in。这种铣刀的侧面呈盘状，有间隙，可以防止粘连。与圆柱形铣刀相比，这种类型的铣刀每英寸直径上的齿数更多，通常用于铣削较深的、狭窄的槽，并可用于切割加工。

练习答案

1. a）多齿刀具 b）金属切削量 c）加工出好的表面粗糙度 d）平面铣削
   e）端面铣削 f）逆铣 g）顺铣 h）心轴铣刀和带柄铣刀

第 8 单元　钻头与钻床

钻头按以下几个特征进行分类：（1）钻头的直径：一般从 0.30mm 至 100mm 不等；（2）钻头的材料：常用碳钢或高速钢；（3）柄部类型：直柄和锥柄两种；（4）钻头长度。

直柄麻花钻只能装在卡盘上使用，并且还要依赖于卡盘与柄部协调一致。锥柄麻花钻在这方面更令人满意。在锥柄尾部有一小段平直的柄脚，与主轴上的孔（槽）相连接。莫氏锥度是锥度的一个国际标准，用于静配合以精确定位，有 1～6 共 6 个号，主要用于各种刀具、刀柄及主轴锥度。当刀柄锥度小于主轴锥度时，二者便可靠一个套筒进行配合，例如，可以用一个 2 号莫氏锥度的主轴与 1 号莫氏锥度的刀柄进行配合。

钻芯厚度从钻尖到容屑槽的根部逐渐增加，以便给予额外的力，但这会影响钻头横刃的长度增加，就像钻头被重磨一样。这部分较长边界需要额外的进给压力，以克服它所承受的阻力。但这个问题可以用打磨边轮、减少钻芯厚度的方法解决。

当出现以下任何低效率的工作状态时，麻花钻应立即被重磨。(1) 需要过多的进给压力，使钻头进行切削；(2) 表明切削边损伤；(3) 施加压力时，钻头出现振动或很尖的噪声。这会引起钻头摩擦代替切削，将迅速导致过热。

根据钻头作业情况有三种基本类型的钻床：台式钻床，立式钻床和摇臂钻床。台式钻床，顾名思义，通过作业者"感觉"钻头的切削作用，如同他的手控制它进行切削。这些机器安装在底座上或地面上。由于这些钻床只能进行轻型应用，所以它们通常有直径1/2的最大钻头。

台式钻床不包括电动机主要有四个部分：头部、主轴、工作台和底座。主轴在通心轴或套筒中转动并上下运动。主轴轴承被台阶式V形槽带轮及滑轮或变速驱动器驱动。

立式钻床与台式钻床非常相似，但它可以应用于更繁重的工作中。它的驱动装置更加强大，并且有许多类型的立式钻床是由齿轮进行驱动的，因此它可以钻两英寸或更大直径的孔。当改变钻床齿轮的转速时，电动机必须停止。操作者可以通过手动手轮或杠杆来从事进给，也可以操纵大功率的进给装置。该装置可以在底座上升高或降低。

摇臂钻床是最通用的钻孔机。它的尺寸由工作台的直径和从主轴中心到工作台外部的摇臂长度决定。它主要用于加工体积和重量较大的工件的孔，由于太重以至于操作者不能改变钻孔的位置。工件被夹在工作台或底座上，通过摇动摇臂及头部沿摇臂运动使钻头在需要的位置锁定。摇臂和头部可以在工作台上升降然后锁定。摇臂钻床可以用来钻小孔及非常大的孔，并且还能够对零件进行镗孔、铰孔、扩孔和锪削。与立式钻床一样，摇臂钻床也有大功率的进给装置和手动进给装置。

练习答案

1. a）台式钻床　　　　　b）立式钻床　　　　　c）摇臂钻床
   d）直柄麻花钻　　　　e）锥柄麻花钻　　　　f）莫氏标准锥度

阅读材料 A　正齿轮和斜齿轮

轮齿是直的且平行于其轴线的齿轮称为正齿轮。正齿轮传动联接副只能用于联接平行轴。平行轴还可用其他类型的齿轮传动联接，一个正齿轮可与不同类型的齿轮相啮合。

为了避免因受热膨胀而引起的卡死现象、方便润滑和补偿不可避免的制造误差，所有的动力传递齿轮必须有侧隙。也就是说在一对啮合齿轮的节圆上，小齿轮的齿槽宽必须稍大于大齿轮的齿厚，反过来也是如此。对于仪表齿轮，侧隙可通过将一个齿轮从中间切开而获得的两个"拼合齿轮"来消除，拼合齿轮的一半与另一半是成比例的。弹簧力能使拼合轮齿占满小齿轮的齿间全部宽度。

斜齿轮具有某些优点。比如当连接平行轴时，其承载能力高于有相同齿数并用相同刀具切削的正齿轮。因为轮齿的重合作用，其传动更加平稳，允许的节线速度比正齿轮高。节线速度是节圆上的线速度。由于轮齿倾斜于旋转轴线，所以斜齿轮会产生轴向推力。如果单独使用这种齿轮，轴向推力则由轴向轴承来承担。轴向推力问题可通过在一个齿坯上加工两对斜齿来解决。依照加工方法，这种齿轮可以是连续齿人字形齿轮或有退刀槽的双斜齿轮。双斜齿轮特别适合于在高速下高效传递动力。

斜齿轮还能用于相互成任何角度的不平行或不交叉的轴间的连接，最常见的连接角度为90°。

阅读材料 B　切削加工的基本方法

我们日常生活中用到的每一件产品都直接或间接地突显了机械加工的重要性。

（a）美国每年在机械加工及相关操作上的投资超过一千亿美元。

（b）制造业中所使用的机床大多数（80%以上）都从事金属切削加工。

这些事实表明在一般制造技术中金属切削的重要性。因此，为更好地应用切削技术，理解切削过程是很重要的。

切削加工的五种基本方法包括：钻削和镗削、车削、刨削、铣削、磨削。五种基本方法的区别在于适用的情况不同。

钻削是用旋转的钻头加工出一个圆孔。镗削则是用一个旋转的、偏置的单刃刀具对钻削出或铸造出的孔进行精加工。在某些镗床上，刀具固定而工件旋转；在另一些镗床上，情况又会反过来。

车床作为回转机械，通常被认为是所有加工设备之父。车削时要切削的金属工件旋转，切削刀具则向工件进给。

用机床刨削金属的过程类似于用木工刨刨木头。本质区别在于当工件在刀具下面来回运动时，刨刀固定在一个位置不动。通常龙门刨床用于加工大工件，甚至有时可加工表面宽达15～20ft、长是宽的两倍的工件。牛头刨床与龙门刨床的区别在于它的工件是固定的而刀具前后移动。

铣削是仅次于车削的应用最广的加工方法。铣削是使工件与具有多刃切削的旋转刀具相接触，从而实现机械加工。针对不同的工件有多种铣床可供选择。铣床加工的某些形状相当简单，如由圆锯生产的缝槽和平面。其他更复杂的形状则是根据给定刀具的切削刃的形状和刀具的运行轨迹而得到的平面与曲面的各种各样的组合。

磨削是通过工件与旋转的砂轮的接触实现工件的加工。这种操作常用于对经过热处理后变得很硬的工件进行精加工，以得到精确尺寸。这是因为磨削能够纠正由热处理产生的变形。近几年，磨削扩展了在金属强力切削方面的应用。

阅读材料 C　机械零件

任何简单的机床，都是由单一元件即通称为机械零件或部件组成的。所以，假如把机床完全拆卸，就可得到像螺母、螺栓、弹簧、齿轮、凸轮和轴等简单零件，它们是所有机器的组合元件。因此，机械零件就是可以执行某种具体功能，而且能与其他零件配合的单一元件。有时候某些特定的元件必须成对地工作，例如螺栓与螺母、键与轴。在其他情况下，一组零件组成一个装配件，如轴承、联轴器与离合器。

机械零件中最常见的是齿轮，它实际上是由轮子和杆组合并带有齿的轮子。在轴套或轴上旋转的齿轮驱动其他齿轮作加速或减速运动，这取决于主动齿轮的齿数。

其他基本机械零件包含轮子和杆。轮子必须装在轴上才可以转动。轮子用联轴器夹紧固定在轴上，轴必须安装在轴承里，由滑轮带或链条与第二根轴相连，并带动第二根轴转动。支撑结构可用螺栓、铆钉或通过焊接固定装配在一起。这些机械零件的正确使用，取决于是否懂得作用于结构上的力和所用材料的强度等相关方面的知识。

单个机械零件的可靠性是评估整台机器使用寿命的基本因素。

很多机械零件是完全标准化的。普通结构和机械部件的最合适的尺寸可以借助测试或实际经验来确定。采用标准化，可以获得实际应用上的一致性和经济上的实惠。然而，并非所有的零件都是标准的。在汽车工业中，只有紧固件、轴承、轴套、传动链和传动带是标准元件。曲轴及连杆不是标准元件。

阅读材料 D　铣　床

大多数铣床是升降台式铣床，由很多零部件和动力装置组成。安装在底座上的床身是其他部件的支撑结构，许多零部件都安装在床身上，包括带有驱动装置的主轴。这种工作台有三个方向互相垂直的进给运动：（1）工作台的垂直运动；（2）工作台的横向进给运动；（3）工作台纵向进给运动。这些运动都可以通过手动或自动进给来实现。大多数情况下，工作台完成快速进给后，还需要返回到原来的位置。

如前所述，当铣床有互相垂直的三个方向的进给运动时，就称为普通升降台式铣床。它分为立式升降台式和卧式升降台式。图ⅡD-1是一台卧式升降台式铣床。这种铣床的顶部安装有横梁，安装在横梁上的挂架用来支撑铣力杆的悬伸端。图ⅡD-2是一台普通的立式升降台式铣床。这类铣床的主轴安装在立铣头上，可以通过手动或自动完成上下进给运动。立式升降台式铣床特别适用于端面铣削，尤其适用于在水平面上孔的加工，因为工作台的进给运动在加工时更为便捷。

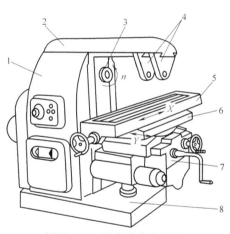
图ⅡD-1　卧式升降台铣床
1—立柱　2—横杆　3—轴　4—心轴支架
5—工作台　6—床鞍　7—升降台　8—底座

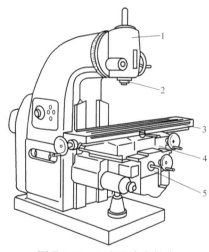
图ⅡD-2　立式升降台铣床
1—主轴　2—心轴　3—工作台
4—床鞍　5—升降台

在万能升降式铣床上，安装有一个万能分度头，它能配合工作台的移动使工件旋转。由于这一特点，工件可以在水平方向旋转，这就增加了铣床的灵活性，使其能进行螺纹铣削，如加工麻花钻、铣刀和斜齿轮。

滑枕式铣床是一种特殊的升降台式铣床。安装在床身上的主轴绕其轴线作旋转运动，用于完成水平方向、垂直方向或任意角度上的铣削，并能灵活地完成各种零件的加工，如刀具和模具及机械加工。

第Ⅲ部分　机电一体化基础

第9单元　电工技术

早在16世纪后期，人们就开始探索电工技术。在电工学中，常用的电气元件有电阻元件、电感元件和电容元件，它们之间的关系是通过电路图或网络来描述的，对电路图的分析可预估实际器件的性能。

纯电阻元件只消耗电能而不储存电能。该特性可以定义为 $R = u(t)/i(t)$。当电压 u 的单位为伏特（V），电流 i 的单位为安培（A），电阻 R 的单位为欧姆（Ω）时，上述公式被称为欧姆定律。

在电场中存储能量的电路元件称为电容器（也称为电容）。当电压在一个周期中变化时，在周期的一段时间内会存储能量，而在随后的时间内又将其释放出来。当去掉电源时，电感元件由于磁场消失而不能保存能量，而电容元件能存储电荷，所以在去掉电源后，电场仍能保持。这种充电的状态将一直保持，直到有放电回路时能量才会释放。电容器中的电荷，$q = Cu$，在电介质中产生电场，这便是电容器存储能量的机理。在简单平板电容器中，一个极板电荷过剩，而另一个极板电荷不足。当电容放电时，两极板上的电荷守恒。电容的符号通常用 C 表示。当电压 u 的单位为 V，电荷 q 的单位为（C），时间 t 的单位为 s 时，电容 C 的单位为（F）。

在磁场中存储能量的电路元件称为电感器（也称为电感）。在时变电流的作用下，电感在一个周期的一段时间里存储能量，而在其他时间段里又释放能量给电源。当电感元件脱离电源时，磁场将不存在，即电感元件如果不连接电源就不能存储能量。在电动机、变压器及类似装置中，有线圈的电路模型都有电感。在频率很高时，甚至对一组平行导体呈现出的电感也必须予以考虑。

在电工技术中，电路是指电流流经的闭合路径。它是为了某种需要，由若干电子设备及元件按照一定方式组成的。它主要由电源、负载、连接导线、控制和保护装置等部分组成。

在电路图中，电源是可以提供能量的设备，它能把其他形式的能转换为电能，常见的电源有电池、发电机等。

负载是各种用电设备的总称，其作用是把电能转换为其他形式的能，如电灯、电扇、电动机等。

连接导线用于把电源和负载连接成闭合回路，从而输送和分配电能，一般常用的导线是铜线和铝线。

控制和保护装置用来控制电路的通断，保护电路的安全，使电路正常工作，如开关、熔断器等。

在电路中，电荷的定向运动称为电流。在金属导体中，电流是电子在外电场作用下有规则地运动形成的。在某些液体或气体中，电流则是正、负离子在电场力作用下向着相反方向的运动形成的。电流是一种物理现象，在数值上等于通过导体横截面的电荷量 q 和通过这些电荷量所用时间的比值。如果在 1s 内通过导体横截面的电荷量是 1C，导体中的电流为 1A。

电流之所以能够在导线中流动，是因为在电流中有着高电势和低电势之分。任意两点间的电势差称为电压，通常用字母 U 表示。电压推动电荷定向移动形成电流。

今天，电子和计算机技术等一系列新兴科学技术蓬勃发展，各种基础科学、应用科学和技术开发之间的知识结构更加紧密，各门学科与专业之间互相渗透、互相交叉，促进了电工学的极大进步，使得电工学可以更好地为人类的生产生活做出更大的贡献。

练习答案

1. a) With time-variable current, the energy is generally stored during some times of the cycle and then returned to the source during others.

b) The capacitor is the circuit element that stores energy in an electric field.

c) The control and protection devices are used to control the on-off of a circuit in order to protect the circuit.

2. 描述　联系　表达　存储　导体　保护　形成　等于

3. a) The electric appliance is a kinds of electrical equipment which can switch the circuit, intermittently or continuously change the circuit parameters, according to the specific signals and external demands for realization of circuit's switching, control, protection, detection, and adjustment.

b) The ordinary conductive materials refer to the metal materials specifically for the current's conduction.

第 10 单元　电子技术

电子技术是 19 世纪末开始发展起来的新兴技术，20 世纪初发展最迅速，应用最广泛，成为近代科学技术发展的一个重要标志。电子技术根据电子学的原理，运用电子器件设计和制造某种特定功能的电路从而解决实际问题。电子技术是对电子信号进行处理的技术，处理的方式主要有：信号的产生、放大、滤波、转换。

第一代电子产品以电子管为核心。19 世纪 40 年代末期世界上诞生了第一只半导体晶体管，它小巧、轻便、省电、寿命长，很快取代了电子管，19 世纪 50 年代末期，世界上出现了第一块集成电路，它把许多晶体管等电子元器件集成在一块硅芯片上，使电子产品向小型发展。集成电路从小规模集成电路迅速发展到大规模集成电路和超大规模集成电路，从而使电子产品向着高效能、低消耗、高精度、高稳定、智能化的方向发展。电子技术通常分为模拟电子技术和数字电子技术。

常规的模拟电子电路主要由电子器件构成，如各种整流电路、放大电路、振荡电路、变换电路等，以及由某些基本功能电路所组成的各种用途的装置或系统。它是构成电子设备的主要电路形式。电子电路在通信、广播、电视、计算机和工业控制等方面得到广泛的应用。它具有传递信息快速、灵敏、精确、容易实现遥控，而且体积小巧、运行可靠、使用方便的优点，是实现现代化先进科学技术的极其重要的一个组成部分。

模拟放大电路的主要特点：

- 处理的信号是连续变化的模拟信号。如音响信号、电视脉冲信号、温度、压力等。
- 晶体管在电路中的作用相当于一个放大器件，而不像在数字电路中晶体管的作用相当于一个开关。
- 分析方法主要采用图解法和微变等效电路法来分析放大电路的静态和动态工作情况。而在数字电路中，经常利用逻辑代数、真值表、卡诺图和状态变换图等来分析输入、输出之间的逻辑关系。

数字信号通常指在时间上和数值上都是离散变化的电信号，其变化都是发生在一系列离散的时刻上，即信号总是在高电平和低电平之间来回变化。数字电路是用来处理数字信号的电子电路，包括对数字信号进行传送、逻辑运算、控制、计数、寄存、显示以及脉冲波形的产生和变换等。

现代数字电路由半导体工艺制成的若干数字集成器件构造而成，它只传输0和1两个状态信息，并做逻辑运算。它的主要特点如下：
- 数字电路的半导体器件工作在开关状态；
- 数字电路的信号是断续变化的脉冲信号；
- 数字电路的分析方法中常用逻辑代数、真值表、卡诺图和状态变换图等。

由于数字电路具有结构简单、容易制造、成本较低、工作可靠等一系列优点，数字电路在自动控制、测量仪器、通信等科学技术领域得到了广泛的应用。

近年来，可编程逻辑器件（PLD）特别是现场可编程门阵列（FPGA）的飞速发展，使数字电子技术开创了新局面，不仅规模大，而且将硬件与软件相结合，使数字电子技术的功能更加完善、使用更加灵活。

练习答案

1. a) It's the high-performance, low consumption, high accuracy, high stability, intelligent direction, and so on.

b) The conventional analog electronic circuit is mainly composed of an electronic device, such as rectifier circuit, amplifier circuit, oscillator circuit, transform circuit, and other basic electronic devices.

c) Its main features are as follows.
- The semiconductor devices of the digital circuits are working in the switch state.
- The signals of the digital circuits are discontinuous changing pulse signals.
- The analysis methods of the digital circuits commonly include the logic algebra, the truth tables, Karnaugh map, the status transformation map, and so on.

2. 转换　结构　制造　交流　可编程的　组合　动态的　远程

3. a) A variety of electronic circuits and systems require a DC power supply. And the DC power supply can be obtained with the transformation of the AC power grid.

b) The amplifying circuit can amplify the weak signals to the desired value.

第11单元　自动控制系统

第二次世界大战时期，自动控制的理论和应用有了重大发展。所谓自动控制系统，是在无人直接参与下可使生产过程或其他过程按期望规律或预定程序进行的控制系统。自动控制系统是实现自动化的主要手段。自动控制系统主要由控制器、被控对象、执行机构和变送器四个环节组成。自动控制系统已被广泛应用于人类社会的各个领域。

冶金、化工、机械制造等工业生产过程中的各种物理量，包括温度、流量、压力、厚度、张力、速度、位置、频率、相位等，都有相应的控制系统。在此基础上不仅建立了控制性能更好和自动化程度更高的数字控制系统，而且建立了具有控制与管理双重功能的过程控制系统。在农业方面的应用包括水位自动控制系统、农业机械的自动操作系统等。过程控制把诸如温度、压力、流量、液位、黏度、密度、成分等过程变量控制为预期值。

现在过程控制方面的许多工作都包含推广使用数字计算机，从而实现过程变量的直接数字控制。目前计算机已经被广泛应用于控制工业过程设备、安全系统、防盗报警、大型建筑物中的中央空调和供暖设备等。

在军事技术方面，自动控制通常应用在各种类型的伺服系统、火力控制系统、制导与控制系统中。在航天、航空和航海方面也常常用到了自动控制系统。除此之外，应用的领域还包括导航系统、遥控系统和各种仿真器。

此外，在办公室自动化、图书管理、交通管理乃至日常生活方面，自动控制技术也都有实际的应用。随着控制理论和控制技术的发展，自动控制系统的应用领域还在不断扩大，几乎涉及生物、医学、生态、经济、社会等所有领域。下面我们讨论现代工业中常见的几个例子。

伺服系统在自动控制领域中是非常常见的。伺服机构，或简称"伺服"，是一种闭环控制系统，其中的被控变量是机械位置或机械运动。该机构的设计使得输出能迅速而精确地响应输入信号的变化。因此，我们可以把伺服机构想象成为一种随动机构。另外还有一种控制输出变化率或输出速度的伺服机构，我们称为速率或速度伺服机构。

在电力工业中，能量的转换与分配非常重要。发电量超过几十万千瓦的现代化大型电厂经常需要复杂的控制系统来表明许多变量的相互关系，并需要提供最佳的发电量。发电厂的控制一般也被认为是一种过程控制的应用，而且通常有多达100个操纵变量受计算机控制。

自动控制已广泛地应用于电能分配。电力系统通常由几个发电厂组成。当负载波动时，电力的生产与传输要受到控制，使该系统达到运行的最低要求。此外，大多数的大型电力系统都是相互联系的，而且系统之间的电能流动也受到控制。

在机械加工过程中，多种工序如镗孔、钻孔、铣削和焊接都必须以很高的精度重复进行。数字控制系统使用称为程序的预定指令来控制一系列工序的运行。完成这些预期工序的指令被编成代码，并且存储在如穿孔纸带、磁带或穿孔卡片等某个介质上。这些指令通常以数字形式存储，故称为数字控制。指令识别要用的工具、加工方法及工具运动的轨迹等参数。

为了向现代化城市的各地区提供大量的运输系统，需要大型、复杂的控制系统。目前正在运行的几条自动运输系统中有每隔几分钟的高速火车。常常需要自动控制来保持火车稳定的速度及提供舒适的加速和停站时的制动。

飞机的飞行控制是在运输领域中的另一项重要应用。由于系统参数的范围广泛以及控制之间的相互影响，飞行控制已被证明为最复杂的控制应用之一。飞机控制系统实质上常常是自适应的，即其操纵本身要适应于周围环境。例如，一架飞机的性能在低空和高空可能是根本不同的，所以控制系统必须作为飞行高度的函数进行修正。船舶转向和颠簸稳定控制与飞行控制相似，但是一般需要更大的功率和较低的响应速度。

练习答案

1. a) The automatic control system is composed of the controller, the controlled object, the implementing agencies, and the transmitter.

b) With the use of the digital computer, automatic control system can process the direct digital control (DDC) of the manipulated variables.

c) The automatic control technology, has the practical application in office automation, library management, traffic management, daily life, and so on.

2. 执行；预定的；由……构成；属于；应用；伺服机构；简称；加速.

3. a) The aircraft autopilot is an automatic device that can maintain or change the flight status. It can stabilize the flight attitude, altitude, and flight path.

b) With the external equipment or device, the automatic control system makes certain operating conditions and the parameters of the machine, equipment or production process automatically modified according to the predetermined rules.

第12单元　机电一体化系统

随着生产和技术的发展，微电子技术、自动化技术不断向机械技术领域内渗透，形成了一门新的学科领域，即机电一体化技术。机电一体化技术一方面极大地提高了产品的性能和市场竞争力；另一方面也大大地提高了产品对环境的适应能力，使人类的活动空间不断扩大至太空，例如美国的阿波罗登月和我国的神舟5号、嫦娥1号、嫦娥2号等都是机电一体化技术发展的结果。由于机电一体化技术对现代工业和技术的发展具有巨大的推动力，引起了世界各国的极大重视。

"机电一体化"（Mechatronics）的英文名词起源于日本，它是取机械学（Mechanics）的前半部分和电子学（Electronics）的后半部分拼成一个新词，表示机械学与电子学两种学科的综合。但是，"机电一体化"并非是机械技术与电子技术的简单叠加，而是把电子技术、信息技术、自动控制功能"揉合"到机械装置中去，通过各种技术的有机结合，使产品的性能达到最佳水平。

随着科学技术的发展，机电一体化产品把机械部分与电子部分有机结合，从系统的观点使其达到最优化。机电一体化的基本概念可概括为：从系统的观点出发，将机械技术、微电子技术、信息技术、控制技术、计算机技术、传感器技术、接口技术等在系统工程的基础上有机地加以综合，实现整个系统最优化而建立起来的一种新的科学技术。

所谓机电一体化，包含机电一体化技术和机电一体化系统两个方面的内容。机电一体化技术是指包括技术基础、技术原理在内的、使机电一体化系统得以实现、使用和发展的技术。机电一体化系统又包括机电一体化产品和机电一体化生产系统。机电一体化生产系统是指运用机电一体化技术把各种机电一体化设备按目标要求组成的一个高生产率、高质量、高可靠性、高柔性、低能耗的生产系统，例如常见的柔性制造系统（FMS）、计算机辅助设计与制造系统（CAD/CAM）、计算机辅助设计工艺（CAPP）和计算机集成制造系统（CIMS）以及各种工业过程控制系统。采用机电一体化技术所制造出来的具有机电一体化特点的新一代产品或设备统称为机电一体化产品。

目前，机电一体化产品及系统已渗透到国民经济和日常工作、生活的各个领域，比如电冰箱、全自动洗衣机、录像机、照相机等家用电器，电子打字机、复印机、传真机等办公自动化设备，工业机器人、自动化物料搬运车、核磁共振成像诊断仪等机械制造设备都属于机电一体化产品。

为了不断满足人们生活的多样化要求和生产的省力、省时和自动化等方面的需要，机电一体化产品不断推陈出新。机电一体化综合利用现代高新技术的优势，在提高精度、增强功能、改善操作性和实用性、提高生产率、降低成本、节约能源、降低消耗、减轻劳动强度、提高安全性和可靠性、改善劳动条件、简化结构、减轻劳动强度、改善劳动条件、提高安全性和可靠性、简化结构、减轻重量增强柔性和智能化程度、降低价格等诸多方面都取得了较为显著的技术效益、经济效益和社会效益,促使着社会和科学技术的进步。总体而言,机电一体化产品具备多功能、高效率、高智能、高可靠性等特点,同时在外观上具有轻、薄、细、小巧的优点,从而在生产生活各个方面都得到了广泛的应用。

机电一体化系统一般由机械部分、传感检测部分、动力及驱动执行部件、控制系统及信息处理等部分组成,这些组成要素内部及其之间,通过接口耦合来实现运动传递、信息控制、能量转换。其中,机械部分为系统的支撑部件;动力系统为系统正常运行提供动力;传感检测部分对系统运行中所需要的本身和外界环境以及各种参数及状态进行检测,将其转换为可识别信号,传输到信息处理单元;控制系统将传感检测部分所检测到的信息以及外部的输入命令进行集中、存储、分析、加工,根据信息处理结果,按照一定的程序和节奏发出相应的指令,控制整个系统有序地进行工作;驱动部件根据控制信息和指令执行相应的动作。机电一体化系统的构成要素使其具备了控制、检测、动力、动作、构造等五大功能,这五个部分各司其职,担任着不同的作用。

练习答案

1. a) The basic concept of mechatronics can be summarized as follows: from the viewpoint of system, the organic synthesis of mechanical technology, microelectronics, information technology, control technology, computer technology, sensor technology, and interface technology in system engineering to realize the whole system optimization, so as to be a new science and technology.

b) The developing direction of the mechatronics is the multi-function, high efficiency, high intelligence, high reliability, and so on.

2. 机电一体化　叠加　结合　学科　高质量　特征　统一的　诊断

3. a) With the development of the science and technology, the mechatronics products and mechatronics technology change rapidly throughout every corner of life.

b. Mechatronics is a comprehensive concept. With the high technical content, the additionally technical values of the mechatronics products improve according to the combinative degree of the mechanism and electronics.

阅读材料 A　交　流　电

交流电在很短的时间内电流强度从零开始,增大到最大值,然后再降到零,然后电流反方向流动,它的强度从零增大到最大值,然后再降到零。电流以这种方式来回地变化。交流电是指周期性改变流动方向的电流。用来表示交流电随时间变化而变化的图称为交流电的波形图。通常周期用 T 来表示,用秒来计量。周期的倒数称为频率,定义为发生在单位时间内的周期数,我们通常用赫兹或周每秒来表示频率。

当交流电的所有数值以半个周期为界限分成大小相等、符号相反的两部分时,交流电流称为对称的。对称交流电流的半个周期的平均值是指电流从零开始经过半个周期的平均值。交流电的平均值很少使用,它只有在交流电整流时才用到。交流电流的方均根或有效值更有用。这个值等于一个周期中瞬时值二次方的平均值的二次方根。不难证明，交流电的有效值等于一直流电的数值，该直流电通过一给定的电阻产生的热量和该交流电通过同一电阻产生的热量相等。

实际上电流和电压的波形与正弦波本质上是有区别的，但在大多数情况下这些波和正弦波十分接近。因而，当我们把波形近似认为正弦波时，所得的结果和实际情况相比有足够的精确度。

阅读材料 B　数 字 电 视

自菲罗特法斯沃茨 1927 年在其印第安纳的车库里组装出第一台电视机以来，电视机的工作方式发生了两个重大的变化：其一是，1954 年引用的彩色；其二是，20 世纪 70 年代由电子管改为晶体管。现在一场深刻的变革即将发生，这就是数字电视。数字电视利用不同的方式进行信号传输，将会大大改变未来电视机的画面及工作方式。

这种数字电视机是介于电子计算机终端设备与现有电视机之间的产物。虽然它将带来的差异也许不是惊人的，但随着现场直播的变焦效果、立体声及定格景观等将成为平常的事，其质量的改进将日益被人们重视。数字电视有希望为电视观众提供比目前市场出售的电视更清晰更调和的画面。

自从 20 世纪 50 年代以来，几乎所有的电子产品的体积都已经逐步地缩小。现在计算器可以放进支票簿内，只有砖块大小的立体扬声器能播放出巨大的声响，激光唱片压缩了唱片工业。可在整个变革过程中，人们的电视机却依然很庞大，这主要是因为，缩小电视机的主要部件之一——显像管的体积缩小比较困难，而且也不经济。

MRS 技术股份有限公司研发出一种平版印刷系统，专门用来制造平版有源矩阵液晶显示器。4500 型平版印刷机（人们给该机起的名字）能够产生 1in 厚的 18in 彩色液晶显示器。它们的重量只有几英镑，可以像油画一样挂在墙壁上。

阅读材料 C　适应性控制系统

适应性控制系统是一种自动调整其参数以补偿过程特性的相应变化的系统。当然，必须有一些作为适应程序依据的准则。为被控制量规定一个数值是不够的，因为要满足这一指标，不仅需要适应性控制，还必须另外规定被控制量的某种"目标函数"。

一个给定过程的目标函数可能是被控制量的衰减度。因而，存在两个回路，一个回路操作被控制量，另一个则操作其衰减度。由于衰减度标志着回路动态增益，因此这种系统被称为动态适应性系统。也有可能为一个过程规定一个静态增益的目标函数，这种控制系统就是静态适应性系统。

动态适应性系统的主要功能是给控制回路一个始终如一的稳定度。因此，动态回路增益就是被控制量的目标函数，其数值要予以规定。凡是能满足目标函数的被控制量的数值是针对该过程的一些主要情况，那么就能够容易地为适应性控制编制出程序。例如，各种空气流量和温度情况下的最佳燃油-空气比可以是已知的。因此，用改变控制器设定值的方法来设计控制系统，使燃油-空气比适应于空气流量与温度的变化，作为动态适应系统例子中的一个流量函数。

动态适应性系统控制回路的动态增益，因此，与之相应的静态适应系统就寻求不变的静态过程增益。这就意味着静态过程增益是变化的,而且有一个特定值是所期望的。

阅读材料 D　新型机电产品展望

随着科学技术的发展,涌现出越来越多的新型机电产品,其中机器人就是一种新兴的智能化的机电产品。当你看到电视上机器人在做家务,你可能没有意识到机器人已经存在于你的生活中;根据定义,洗衣机、电热炉等全部属于机器人。也能够设计出在实验室和外层太空研究中做危险性工作的机器人。

所有发射到外层太空的卫星上都安装有机器人,这些机器人通过无线电把有关宇宙的一些重要资料如温度、辐射等发回给地球上的主人。从太空很高的位置上它们甚至能够拍摄到地球和其他行星的相片。

当第一架太空船在火星和金星上着陆时,它上面带的是机器人而不是人类。机器人能够描绘出这些星体的表面,开展必要的地理研究,开发未知的地方。在过去的20年里,工业界认识到:为了提高在世界市场的竞争力,必须提高生产率,降低生产成本。因为熟练技术工人的数量逐渐减少,以及越来越少的人员愿意从事单调、繁重、环境恶劣的工作,工业界发现很有必要对大部分生产过程实现自动化。计算机的发展使工业界生产可靠的机床和机器人成为可能。这些机床和机器人能使制造过程生产力更高、可靠性更好,从而提高产品在世界市场上的竞争力。

第Ⅳ部分　机电一体化技术

第 13 单元　AT89S51 介绍

产品概述

AT89S51 是一款低功耗、高性能的 CMOS 型 8 位单片机,片内含 4KB 的在系统可编程序的 Flash 只读存储器,器件采用 ATMEL 公司的高密度非易失性存储技术生产,兼容工业标准 80C51 的指令系统及引脚。片内 Flash 程序存储器既可在线编程,也可用传统非易失性存储器编程器。在单片芯片中集成了 8 位通用 CPU 和在系统可编程序的 Flash 只读存储器,ATMEL 公司的功能强大、低价位 AT89S51 单片机可为您提供许多高性价比的应用场合,可灵活应用于各种控制领域。

性能参数

- 与 MCS-51 系列产品兼容
- 4KB 在系统编程(ISP)Flash 闪速存储器,达 1000 次擦写周期
- 4.0~5.5V 工作电压范围
- 全静态工作模式 0~33MHz
- 三级程序加密锁
- 128×8 位内部 RAM
- 32 个可编程 I/O 口线
- 2 个 16 位定时/计数器
- 6 个中断源
- 全双工串行 UART 通道

- 低功耗空闲和掉电模式
- 从空闲模式唤醒中断系统
- 看门狗
- 双数据指针
- 掉电标识
- 快速编程
- 灵活的在系统编程（字节和页写模式）

概述

AT89S51 提供以下标准功能：4KB Flash 闪速存储器，128B 内部 RAM，32 个 I/O 口线，看门狗（WDT），两个数据指针，两个 16 位定时/计数器，一个 5 向量两级中断结构，一个全双工串行通信口，片内振荡器及时钟电路。同时，AT89S51 可降至 0Hz 的静态逻辑操作，并支持两种软件可选的节电工作模式。空闲方式停止 CPU 的工作，但允许 RAM、定时/计数器、串行通信口及中断系统继续工作。掉电方式保存 RAM 中的内容，但振荡器停止工作并禁止其他所有部件工作直到下一个外部中断或硬件复位。

引脚配置

引脚封装有双列直插式封装（PDIP）、带引线的塑料芯片载体（PLCC）表面贴装型封装，如图 13-1a、b 所示。

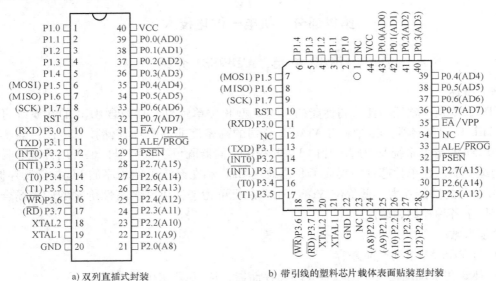

a) 双列直插式封装　　　　b) 带引线的塑料芯片载体表面贴装型封装

图 13-1　双列直插式封装和带引线的塑料芯片载体表面贴装型封装引脚配置

引脚功能说明

VCC　　　　　　　电源电压

GND　　　　　　　地

Port 0

P0 口是一组 8 位漏极开路型双向 I/O 口，也是地址/数据总线复用口。作为输出口用时，每位能驱动 8 个 TTL 逻辑门电路，对端口写"1"可作为高阻抗输入端用。

在访问外部数据存储器或程序存储器时，P0 口线分时转换成地址总线（低 8 位）和数据总线复用，在访问期间激活内部上拉电阻。

在 Flash 编程时，P0 口接收指令字节，而在程序校验时，输出指令字节，在程序校验时，要求外接上拉电阻。

Port 1

P1 口是一个带内部上拉电阻的 8 位双向 I/O 口，P1 口的输出缓冲级可驱动（吸收或输出电流）4 个 TTL 逻辑门电路。对端口写"1"，通过内部的上拉电阻把端口拉到高电平，此时可作为输入口。作为输入口使用时，因为内部存在上拉电阻，某个引脚被外部信号拉低时会输出一个电流 I_{IL}。在 Flash 编程和程序校验时，P1 口接收低位地址。

端口引脚	第二功能
P1.5	主出从入（用于 ISP 编程）
P1.6	主入从出（用于 ISP 编程）
P1.7	串行移位时钟 SCK（用于 ISP 编程）

Port 2

P2 口是一个带内部上拉电阻的 8 位双向 I/O 口，P2 口的输出缓冲级可驱动（吸收或输出电流）4 个 TTL 逻辑门电路。对端口写"1"，通过内部的上拉电阻把端口拉到高电平，此时可作为输入口。作为输入口使用时，因为内部存在上拉电阻，某个引脚被外部信号拉低时会输出一个电流 I_{IL}。

在访问外部程序存储器或 16 位地址的外部数据存储器（例如执行 MOVX @ DPTR 指令）时，P2 口送出高 8 位地址数据。此时，P2 口利用强大的内部上拉发送高电平。在访问 8 位地址的外部数据存储器（如执行 MOVX @ Ri 指令）时，P2 口用于发送特殊功能寄存器 P2 的内容。

Flash 编程或校验时，P2 口也接收高位地址和其他控制信号。

Port 3

P3 口是一组带有内部上拉电阻的 8 位双向 I/O 口。P3 口输出缓冲级可驱动（吸收或输出电流）4 个 TTL 逻辑门电路。对 P3 口写入"1"时，它们被内部上拉电阻拉高并可作为输入端口。作为输入端时，被外部拉低的 P3 口将用上拉电阻输出电流（I_{IL}）。

P3 口还接收一些用于 Flash 闪速存储器编程和程序校验的控制信号。

P3 口除了作为一般的 I/O 口线外，更重要的用途是它的第二功能，如下表所示。

端口引脚	第二功能
P3.0	RXD（串行输入口）
P3.1	TXD（串行输出口）
P3.2	$\overline{INT0}$（外部中断 0）
P3.3	$\overline{INT1}$（外部中断 1）
P3.4	T0（定时/计数器 0）
P3.5	T1（定时/计数器 1）

端口引脚	第二功能
P3.6	$\overline{WR}$（外部数据存储器写选通）
P3.7	$\overline{RD}$（外部数据存储器读选通）

RST

复位输入。当振荡器工作时，RST 引脚出现两个机器周期以上高电平将使单片机复位。看门狗溢出将使该引脚输出 98 个晶振周期的高电平。设置 SFR AUXR 的 DISRTO 位（地址 8EH）可打开或关闭该功能。DISRTO 位缺省为 RESET 输出高电平打开状态。

ALE/PROG

当访问外部程序存储器或数据存储器时，ALE（地址锁存允许）输出脉冲用于锁存地址的低 8 位字节。对 Flash 存储器编程期间，该引脚还用于输入编程脉冲（PROG）。

即使不访问外部存储器，ALE 仍以时钟振荡频率的 1/6 输出固定的正脉冲信号，因此它可对外输出时钟或用于定时目的。要注意的是：每当访问外部数据存储器时将跳过一个 ALE 脉冲。

如有必要，可通过对特殊功能寄存器（SFR）区中的 8EH 单元的 D0 位置位，可禁止 ALE 操作。该位置位后，只有使用一条 MOVX 或 MOVC 指令，ALE 才会被激活。此外，该引脚会被微弱拉高。单片机执行外部程序时，应设置 ALE 无效。

PSEN

程序储存允许（$\overline{PSEN}$）输出是外部程序存储器的读选通信号，当 AT89S51 由外部程序存储器取指令（或数据）时，每个机器周期两次 $\overline{PSEN}$ 有效，即输出两个脉冲。当访问外部数据存储器时，没有两次有效的 $\overline{PSEN}$ 信号。

$\overline{EA}$/VPP

外部访问允许。欲使 CPU 仅访问外部程序存储器（地址为 0000H ~ FFFFH），$\overline{EA}$ 端必须保持低电平（接地）。须注意的是：如果加密位 LB1 被编程，复位时内部会锁存 $\overline{EA}$ 端状态。如 $\overline{EA}$ 端是高电平（接 VCC 端），CPU 则执行内部程序存储器中的指令。

Flash 存储器编程时，该引脚加上 +12 的编程电压 VPP。

XTAL1

振荡器反相放大器及内部时钟发生器的输入端。

XTAL2

振荡器反相放大器的输出端。

练习答案

1. a) AT89S51 是 Atmel 公司生产的一个 8 位单片机。是
   b) AT89S51 有 4 个双向 I/O 口和 4K 字节在系统可编程程序存储器。是
2. 请介绍 AT89S51 单片机的组成。

The AT89S51 provides the following standard features: 4K bytes of Flash, 128 bytes of RAM, 32 I/O lines, watchdog timer, two data pointers, two 16-bit timer/counters, a five-vector two-level interrupt architecture, a full duplex serial port, on-chip oscillator, and clock circuitry.

3. a) AT89S51 具有 128B 的内部 RAM，这 128B 可利用直接或间接寻址方式访问，堆栈操作可利用间接寻址方式进行，128B 均可设置为堆栈区空间。

b) WDT 是为了解决 CPU 程序运行时可能进入混乱或死循环而设置，它由一个 14bit 计数器和看门狗复位 SFR（WDTRST）构成。外部复位时，WDT 默认为关闭状态，要打开 WDT，用户必须按顺序将 01EH 和 0E1H 写到 WDTRST 寄存器（SFR 地址为 0A6H），当启动了 WDT，它会随晶体振荡器在每个机器周期计数，除硬件复位或 WDT 溢出复位外没有其他方法关闭 WDT，当 WDT 溢出时，将使 RST 引脚输出高电平的复位脉冲。

第 14 单元　西门子可编程序控制器介绍

产品概述

S7-200 系列是一种可编程序逻辑控制器（Micro PLCs）。它能够控制各种设备以满足自动化控制需求。

S7-200 的用户程序中包括了位逻辑、计数器、定时器、复杂数学运算以及与其他智能模块通信等指令内容，从而使它能够监视输入状态，改变输出状态以达到控制目的。紧凑的结构、灵活的配置和强大的指令集使 S7-200 成为各种控制应用的理想解决方案。

新内容

SIMATIC S7-200 的新特性包括下列内容。表 14-1 列出了支持这些新特性的 S7-200 CPU。

表 14-1　S7-200 CPU

S7-200 CPU	订货号
CPU 221 DC/DC/DC 6 输入/4 输出	6ES7 211-0AA23-0XB0
CPU 221 AC/DC/继电器 6 输入/4 继电器输出	6ES7 211-0BA23-0XB0
CPU 222 DC/DC/DC 8 输入/6 输出	6ES7 212-1AB23-0XB0
CPU 222 AC/DC/继电器 8 输入/6 继电器输出	6ES7 212-1BB23-0XB0
CPU 224 DC/DC/DC 14 输入/10 输出	6ES7 214-1AD23-0XB0
CPU 224 AC/DC/继电器 14 输入/10 继电器输出	6ES7 214-1BD23-0XB0
CPU 224XP DC/DC/DC 14 输入/10 输出	6ES7 214-2AD23-0XB0
CPU 224XP AC/DC/继电器 14 输入/10 继电器输出	6ES7 214-2BD23-0XB0
CPU 226 DC/DC/DC 24 输入/16 输出	6ES7 216-2AD23-0XB0
CPU 226 AC/DC/继电器 24 输入/16 继电器输出	6ES7 216-2BD23-0XB0

-S7-200 CPU 包括 CPU 221、CPU 222、CPU 224、CPU 224XP 和 CPU 226 等型号，它们有如下特性。

新 CPU 硬件支持：通过关闭在运行模式下编辑程序的可选功能来获取更多的程序存储区。CPU 224XP 支持集成的模拟量 I/O 和两个通信端口。CPU 226 带有附加的输入滤波器和脉冲捕获功能。

- 新型存储卡支持：S7-200 资源浏览器的使用、存储卡的转换、比较以及编程选择。

- STEP 7-Micro/WIN 4.0 版是用于 S7-200 的 32 位编程软件包，它包括支持最新 CPU 增强功能的新软件工具和改进过的软件工具：PID 自动整定控制面板、PLC 内置位置控制向导、数据归档向导和配方向导。

新的诊断工具：可组态诊断 LED。

新指令：夏令时（READ_RTCX 和 SET_RTCX）、间隔定时器（BITIM，CITIM）、清除中断事件（CLR_EVNT）以及诊断 LED（DIAG_LED）。

POU 和库的增强功能：新型字符串常量和添加的间接寻址支持更多存储类型，增强了使用 USS 库函数对西门子变频器（master drives）读写参数功能的支持。

改进的数据块：数据块页、数据块自动增量。

改进的 STEP 7-Micro/WIN 可用性。

S7-200 CPU

S7-200 CPU 将一个微处理器、一个集成电源输入回路和输出回路集成在一个紧凑的封装中，从而形成了一个功能强大的微型 PLC，如图 14-1 所示。在下载了程序之后，S7-200 将保留所需的逻辑，用于监控应用程序中的输入输出设备。

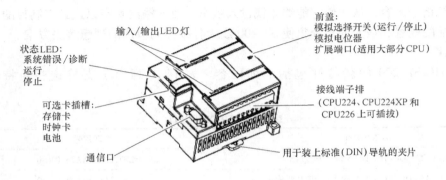

图 14-1　S7-200 小型 PLC

西门子公司提供多种类型的 CPU，帮助你生成适用于各种应用的有效解决方案。

S7-200 扩展模块

为了更好地满足应用要求，S7-200 系列为您提供多种类型的扩展模块。您可以利用这些扩展模块给 S7-200CPU 增加其他功能，比如数字量模块、模拟量模块、智能模块以及其他扩展模块。

STEP 7-Micro/WIN 编程软件

STEP 7-Micro/WIN 编程软件为用户开发、编辑和监控自己的应用程序提供了良好的编程环境。为了能方便高效地开发应用程序，STEP 7-Micro/WIN 软件提供了三种程序编辑器。为了便于您找到所需的信息，STEP 7-Micro/WIN 提供了一个详尽的在线帮助系统以及一个文档光盘，该光盘含有帮助手册的电子版、应用示例和其他有用的信息。

计算机配置要求

STEP 7-Micro/WIN 既可以在 PC 上运行，也可以在西门子公司的编程器上运行，比如 PG 760，如图 14-2 所示。PC 或编程器需满足如下最小配置：

- 操作系统为 Windows 2000 和 Windows XP（专业版或家庭版）。
- 至少 100MB 的可用硬盘空间。

Appendix | 145

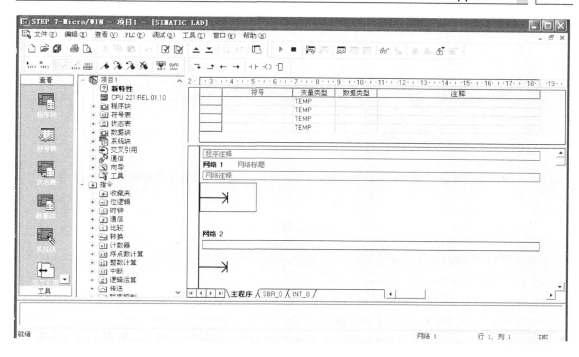

图 14-2　STEP 7-Micro/WIN

- 鼠标（推荐）。

安装 STEP 7-Micro/WIN

将 STEP 7-Micro/WIN 的安装光盘插入计算机的 CD-ROM 中，安装向导程序将自动启动并引导您完成整个安装过程。

通信方式选择

您可以有两种方式连接 S7-200 和您的计算机：通过 PPI 多主站电缆直接连接，或者通过带有 MPI 电缆的通信处理器（CP）卡连接。PPI 多主站编程电缆是连接计算机和 S7-200 的最常用和最经济的方式。它将 S7-200 的编程口与计算机的串行通信口相连。PPI 多主站编程电缆也能用来将其他通信设备连接至 S7-200。

显示面板

文本显示器（TD 200 和 TD 200C）

TD 200 和 TD 200C 是 20 字符双行显示器，可以连接在 S7-200 上，如图 14-3 所示。通过 TD 200 向导，您可以轻松地对 S7-200 编程，实现文本消息和其他应用程序数据的显示。TD 200 和 TD 200C 可以为您提供价格低廉的人机界面，通过它们您将能够查看、监控和改变应用程序的过程变量。

TP070 和 TP170 小型触摸屏

TP070 和 TP170 小型触摸显示屏可以连接至 S7-200，如图 14-4 所示。通过该触摸屏，您可以自定义操作界面。通过触摸屏，这些设备可以显示自定义图形、滚动条、应用程序变量和用户自定义按钮等。

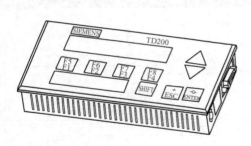 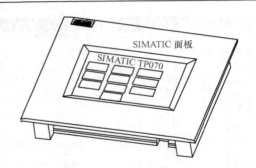

图 14-3 文本显示单元（TD 200 and TD 200C）　　　图 14-4 触摸屏单元

练习答案

1. a) STEP 7-Micro/WIN 4.0.

   b) PPI Multi-Master programming cable.

   c) Insert the STEP 7-Micro/WIN CD into the CD-ROM drive of your computer. The installation wizard starts automatically and prompts you through the installation process.

2. Digital inputs: Each scan cycle begins by reading the current value of the digital inputs and then writing these values to the process-image input register.

3. 模拟量输入：除非使能模拟量滤波，否则 S7-200 在扫描周期中不会刷新模拟量输入值。模拟量滤波会使您得到较稳定的信号。您可以使能每个模拟量输入通道的滤波功能。

第 15 单元　液压技术

　　泵的作用是给液压油加压，换句话说，就是给机器提供动力。阀门的作用是对液体的流向、压力高低、流量大小进行控制以满足要求。用尽可能最简单直白的方式来说明它们在"回路"中的作用，那就是压力油从油箱中流经泵、阀门、驱动单元后再回到油箱，其参考工作原理如图 15-1 和图 15-2 所示。

　　首先，从图 15-1 可以知道液压回路的总体概念（略去了进给机构），然后从图 15-2 可以清楚地了解溢流阀和换向阀的操作，从图 15-2 也不难理解包括进给机构的液压回路的工作细节。图 15-1 所示的溢流阀打开（调速活塞打开），允许通过 V 端口（9）向油箱中排液体，机器工作。压力油从油箱沿箭头的方向通过 R1 经吸入口（1），从油箱右手边的液压缸的活塞口（3）出去，并推动活塞（工作台）向左运动。从而推动活塞左边的液压油通过口（4），穿过端口（2）以及通过 V 端口（9）到油箱。阀门换向，油通过阀门反向流动（如图 15-2），活塞的运动方向也改变。如图 15-2a 所示（是图 15-1 中阀门的放大），压力油从油箱和泵，通过 R1 到（1）到（3）到液压缸，推动活塞向左运动；油缸中活塞另一边的压力油穿过（4）至（2）到（9）通过 R2 返回到油箱。注意液压油能进入（1），但不能进入（5），因为它被（11）阻挡；也不能穿过（12），因为它被（13）阻挡。

　　在图 15-2b 中阀门反转到左边，这只是由于关闭（1）同时打开（5）。压力油在液压缸中通过 R1 到（5）再到（4）流到左边，与此同时，缸右边的压力油通过（3）至（2）到（9）到 R2 被排回到油箱。注意：和图 15-2a 一样，压力油的流动方式如前所述。除此

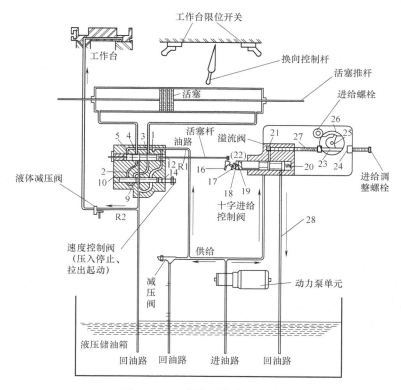

图 15-1　一种典型的液压回路

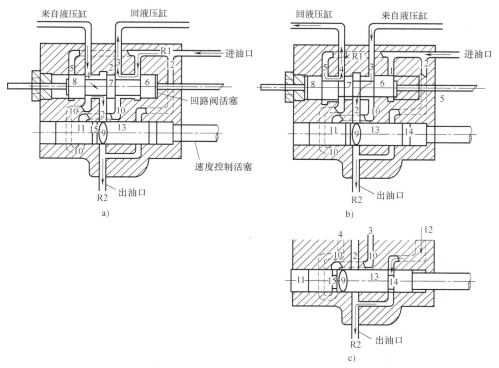

图 15-2　图 15-1 中控制阀的放大图

之外，压力可以通过手动调节柱塞来改变。V（9）口是一个切口，通过旋转柱塞口调节端口的大小，当然，油流过端口，通过驱动液压缸中的活塞，从而对工作台进行驱动。

当阀门如图 15-2c 所示那样被打开时，V（9）口无油排出，工作台保持静止状态，压力油通过管线（12）和（14）从旁路流走。这样可以停止工作台动作，而不是完全关闭（9），因为它可以避免液压油通过减压阀。如图 15-2 所示，在活塞被推进去后，空间（15）打开管线（10），压力油通过（10）在缸体的任何一边流动，实现对工作台的控制。随着工作台来回手动进给，管线中的油通过活塞被从一侧推到另一侧。要理解这些并不难，先参考图 15-2c，注意从（10）到（15）的管线是通畅的，然后看图 15-2a，想象一下，这条路线和图 15-2c 一样也是开放的。然后，当工作台从右到向左移动时，液压油将在缸体内通过（4）到（10）通过（15）到（10）到（3）流动。当反向后，工作台由左到右移动，油将会逆流，由（3）到（10）通过（15）到（10）至（4），最后到缸的左边。

每一个液压系统使用一个或多个泵给油加压。液压油内存储压力，反过来，使液压系统输出单元对外执行工作，因此，压力油可用于在缸体中推动活塞或驱动液压马达中的轴转动。目前在液压系统中常使用 100Pa 或少于 100Pa 的压力。但是也有需要高于 1000Pa 的情况，因此我们发现每个现代液压系统至少使用一个泵进行加压。

液压系统常用三种类型的泵：1）旋转的，2）往复式的和 3）离心泵

简单液压系统可以使用一种类型的泵，所用泵的类型和具体任务特性有关。根据不同液压系统的特点和要求匹配泵，通常联合使用两种类型的泵。例如，一个离心泵可用于给往复泵加压，或者旋转泵可以为往复式泵根据相关情况提供压力油。

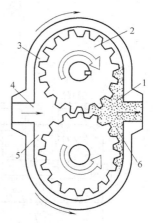

图 15-3　旋转齿轮泵
1—排放口　2—驱动齿轮
3—压力油在油箱中被齿轮
以钱伯斯形式挤压运转
4—吸入口　5—齿轮的相互啮合产生
真空，压力油从油箱中进入
6—齿轮再次啮合迫使油从压力孔排出

这些都是建立在许多不同的设计基础上的，是非常流行的现代液压系统。目前最常见的旋转泵是齿轮泵和螺杆泵。每种泵都有各自的优势以满足最适合给定的应用。如图 15-3 所示的齿轮泵，在密合的壳体中有两个啮合的齿轮，一个主动齿轮旋转，带动另一个从动齿轮旋转。传动轴通常与上面的齿轮连接。当泵工作时，通常是齿轮旋转使气体排出壳体，从而使泵壳产生局部真空，使液压油抽吸到泵的一面，大气压力从外部迫使液压油进入泵口。这时的流体被困在上部和下部齿轮之间，齿轮继续旋转，液压油通过齿轮泵排出。

液压油压力上升是由泵的啮合轮齿和套管的挤压作用而产生的。腔内真空的形成，造成更多的流体被吸取进入泵。一个齿轮泵是定量的，在给定轴的速度时其流量是不变的。只有通过改变转速来调节泵的流量。现代齿轮泵的最高压力可达 3000Pa，齿轮泵有许多叶片，可以在转子的槽内自由地滑入滑出。当转子由泵驱动旋转时，离心力、弹簧或流体压力使叶片向外移动，并与泵壳内孔或凸轮相互作用。随着转子旋转，流体在叶片间流动并通过吸入口。液体在泵壳内部流动，直到到达出口进入排放管道。

压力控制阀用于在液压回路中维持不同部分的压力水平。压力控制阀通过两种方式控制

压力水平：(1) 将高压区的液体转移到低压区，从而限制高压区的压力，或（2）限制液体流动到其他区域。阀门按用途可以分为：安全阀、压力阀、流量阀、反向阀、溢流阀等。限制液体流到另一个区域的阀类可能是减压型的。压力控制阀也可以被定义为常闭式或常开式二通阀。安全阀、溢流阀等通常是常关的，根据执行的设计功能双向阀部分或完全开启。

液压驱动系统应用于机床刀具已不是新事，然而大规模使用是最近才发生的。实际上，现代整体泵的出现，促进了这种机床操作形式的发展。液压驱动机床刀具具有很多的优势，其中之一是它可以在广阔的范围内提供无级变速。除此之外，还可以像改变速度一样方便地改变方向，柔性和韧性是液压系统的另一个优点，一个大的改进是液压工具可以不损伤表面光洁度并且实现重切削，减少损伤。到目前为止，很大比例的机床刀具驱动局限于直线运动，一个转子泵被用于驱动一个或多个线性液压马达，通常是活塞式。某些情况下，某个液压车床中刀具的直线运动和工件的旋转运动可能是液压驱动/或控制，旋转运动是使用旋转液压马达来实现的。

练习答案

1. a) 低速大转矩，结构紧密，稳定性好，滑动平稳，速度和方向控制灵活，可实现无级化输出。

b) 一般来说，液压传动可以分为直线和螺旋的，螺旋式传动产生螺旋运动，而由活塞及缸体部件产生往复的运动是线性运动。

2. a) In general, hydraulic drives may be divided into rotary and linear types.

b) When oil is fed to hydraulic motor or hydraulic cylinder, it is converted into mechanical energy.

c) There is little constructional difference between hydraulic motor and pump. Any pump may be used as a motor.

第 16 单元　气动技术

气动技术是以压缩机为动力源，以压缩空气为工作介质，进行能量传递或者信号传递的工程技术。压缩空气经过一系列控制元件后，将能量传递至执行元件，输出力（直线气缸）或者力矩（摆缸或气马达）。气动技术是研究压缩空气流动规律的科学技术。

与液压、电气技术相比，气动技术是实现低成本自动化的最佳手段。气动元件结构简单、紧凑、易于制造、价格便宜、介质取之不尽、对环境污染较少、可靠性高、使用寿命长，由于空气的可压缩性，因此可以实现能量储存并进行远距离传输。

与机械、液压和电气技术相比，气动技术的工作适应性更广，易于实现快速的直线往复运动、摆动、高速移动和输出力。气动技术的运动速度调节方便，改变运动方向简单，在恶劣环境下工作安全可靠，容易实现防潮、防爆，安装与控制有较高的自由度，具有过载保护能力。与机械、液压、电气技术相比，气动技术也有一定的缺点：由于空气的压缩性，气缸的动作速度容易受到负载的影响，很难达到匀速状态，运动速度低时，摩擦力占推力的比例较大，易出现爬行，因此低速稳定性不好，气缸的输出力相对较小。

气动系统由下面几种元件及装置组成。

气源装置：压缩空气的发生装置以及压缩空气的存储、净化的辅助装置。它为系统提供合乎质量要求的压缩空气。执行元件：将气体压力能转换成机械能并完成做功动作的元件，如气缸、气马达。控制元件：控制气体压力、流量及运动方向的元件，如各种阀类；能完成一定逻辑功能的元件，如气动传感器及信号处理装置。气动辅助元件：气动系统中的辅助元件，如消声器、管道、接头等。

气动技术，全称气压传动与控制技术，是生产过程自动化和机械化的最有效手段之一，具有高速高效、清洁安全、低成本、易维护等优点，被广泛应用于轻工机械领域中，在食品包装及生产过程中也正在发挥越来越重要的作用。

气动技术应用的最典型的代表是工业机器人。它代替人类的手腕、手以及手指，能正确并迅速地做抓取或放开等细微的动作。除了工业生产上的应用之外，在游乐场的过山车上的刹车装置，机械制作的动物表演以及人形报时钟的内部，均采用了气动技术，实现细小的动作。

液压可以得到巨大的输出力但灵敏度不够；另一方面要用电能来驱动物体，总需要用一些齿轮，同时不能忽视漏电所带来的危险。而与此相比，使用气动技术既安全又对周围环境无污染，即使在很小的空间里，也可以实现细小的动作。尺寸相同时，其功率能超过电气设备。在生产线上，实现前进、停止、转动等细小简单的动作，在自动化设备中不可或缺。气动技术也能应用于其他方面，如制造硅晶片生产线上不可缺少的电阻液涂抹工序中使用的定量输出泵以及与此相配合的周边机器。

另外，虽然气动技术在各工业部门已经获得了广泛应用，但是，在许多应用之间还是存在着相当大差异的。就应用气动技术来说，最基本条件就是要有一台空气压缩机，对已有用于其他用途的空气压缩机的地方，应用气动技术就更加方便。特别是在一些非生产加工部门，如畜牧业、种植业或服装业，情况更是如此。在机器设备制造领域中，大多数场合都有空气压缩机，且气动技术已有应用，每个应用项目在本质上也有许多相似之处，因此我们可以对机器设备制造中的气动技术应用情况进行归纳总结。

1. 农业

（1）田间作业　田间工作设备的倾斜、提升和旋转装置；农作物保护和杂草控制设备；包装袋提升机和其他搬运设备。

（2）动物饲养　饲料计量和传送；粪便收集和清除；蛋类分选系统；通风设备；剪羊毛和屠宰。

（3）动物饲料生产　动物饲料生产和包装的装卸设备；计量装置；搅拌和称量系统；存储、计数和监测装置。

（4）林业　林业是指保护生态环境保持生态平衡，培育和保护森林以取得木材和其他产品，利用林木的自然特性以发挥防护作用的生产部门。

（5）种植业　温室通风设备；收割机；水果和蔬菜分选设备。

2. 公共设施

（1）热电站　锅炉房通风设备；遥控阀；气动开关控制。

（2）核电站　燃料和吸收器进给装置；进料口和手控气锁之间的互锁；测试和计量装置；自动操作系统。

（3）供水系统　水位控制；遥控阀；下水道和废水处理中的耙式齿轮滤网和控制阀的操作。

3. 采矿

在露天和地下矿场中直接或间接开采矿石的辅助设备。

4. 化学工业

容器盖密封;计量系统;(混合物)搅拌杆调节;实验室中化学物混合装置;浸入电解槽中的升降装置;控制阀操作,称量装置控制,包装,制模;水位控制和过程控制系统。

5. 石油工业

与化学工业相似,用于工厂和实验室的辅助设备。

6. 塑料橡胶工业

(1) 塑料生产　大批量材料传送和分发的控制系统;控制阀操作系统;料箱1门。

(2) 塑料加工　切断机操作;塑模闭合;成形、模压和焊机的关闭装置。

(3) 橡胶加工　安全装置;用于控制和驱动的传送和生产装置;混合器和硫化压机中的关闭装置;测试设备。

练习答案

1. a) Pneumatic technology is the best mean to implement low cost automation compared with hydraulic, electric technologies. Pneumatic element has a simple structure, compact, manufacturing easy, low price, medium inexhaustible, less pollution to the environment, high reliability, and long life, which can stored the energy and can be transmitted over a long distance, because of the air compressibility.

b) Agriculture: (1) field operations, (2) animal breeding, (3) animal feed production, (4) forestry, (5) planting;

Public facilities: (1) thermal power plant, (2) nuclear power plant, (3) water supply system;

Mining;

Chemical industry;

The oil industry;

Plastic rubber industry: (1) plastic production, (2) plastics processing, (3) rubber processing.

c) Pneumatic technology also has certain disadvantages: compared with machinery, hydraulic, and electric technologies, it difficult to reach the uniform state, because of the air compressibility, air cylinder movement speed vulnerable to the influence of the load, low speed movement, friction larger proportion of thrust, creep phenomenon easily, low speed poor stability, and the output of the cylinder force smaller.

2. a) Pneumatic transmission is not smooth like hydraulic transmission, due to the low working pressure (general less than 1 Mpa), the compressibility, and small transfer power.

b) Pneumatic system is a system to transfer power or signal based on gas (commonly used compressed air).

第 17 单元　电气工程

电气工程（Electrical engineering 简称 EE）是现代科技领域中的核心学科之一，更是当今高新技术领域中不可或缺的关键学科。例如正是电子技术的巨大进步才推动了以计算机网络为基础的信息时代的到来，并将改变人类的生活工作方式。电气工程的发展前景同样很有潜力，使得当今的学生就业率一直很高。

从某种意义上讲，电气工程的发达程度代表着国家的科技进步水平。正因为此，电气工程的教育和科研一直在发达国家大学中占据十分重要的地位。

美国大学电气工程学科的机构名称，有的学校称电气工程系，有的称为电气工程与信息科学系，有的称为电气工程与计算机科学系等。该学科（系）在科研、教学及学术组织形式上与国内电气工程学科有较大不同。了解国外学科状态及教学、科研方向，对调整我们的学科方向，提高教学、科研水平具有十分重要的作用。

传统的电气工程定义为用于创造产生电气与电子系统的有关学科的总和。此定义本已经十分宽泛，但随着科学技术的飞速发展，21 世纪的电气工程概念已经远远超出上述定义的范畴。斯坦福大学教授指出：今天的电气工程涵盖了几乎所有与电子、光子有关的工程行为。本领域知识宽度的巨大增长，要求我们重新检查甚至重新构造电气工程的学科方向、课程设置及其内容，以便使电气工程学科能有效地回应学生的需求、社会的需求、科技的进步和动态的科研环境。

今后若干年内，对电气工程发展影响最大的主要因素包括以下几个方面。

1. 信息技术的决定性影响

信息技术广泛地定义为包括计算机、世界范围的高速宽带计算机网络及通信系统，用于传感、处理、存储和显示各种信息的相关支持技术的综合。信息技术对电气工程的发展具有特别大的支配性影响。信息技术持续以指数速度增长，在很大程度上取决于电气工程中众多学科领域的持续技术创新。反过来，信息技术的进步又为电气工程领域的技术创新提供了更新更先进的工具基础。

2. 与物理科学的相互交叉面拓宽

由于晶体管的发明和大规模集成电路制造技术的发展，固体电子学在 20 世纪的后 50 年对电气工程的成长起到了巨大的推动作用。电气工程与物理科学间的紧密联系与交叉仍然是今后电气工程学科的关键，并且将拓宽到生物系统、光子学、微机电系统（MEMS）。21 世纪中的某些最重要的新装置、新系统和新技术将来自上述领域。

3. 快速变化

技术的飞速进步和分析方法、设计方法的日新月异，使得我们必须每隔几年对工程问题的过去解决方案重新全面思考或审查。这对我们如何聘用新的教授，如何培养我们的学生有很大影响。

电力是发展生产和提高人类生活水平的重要物质基础，电力的应用在不断深化和发展，电气自动化是国民经济和人民生活现代化的重要标志。就目前国际水平而言，在今后相当长的时期内，电力的需求将不断增长，电气工程及其自动化科技工作者的需求量将呈上升态势。电气自动化用于工业控制系统，确保设备正常运行并生产出合格的产品，现代工业不是全人工，靠人来操作，而是由机器来操作。启动机器，机器将自动运行，就是电气自动化。

电气自动化，就是利用继电器、传感器等电气元件实现顺序控制和时间控制。其他如一些仪表或伺服电动机，能根据外界环境的变化反馈到内部，从而改变输出量，达到稳定的目的。

4. 电气工程专业人才培养目标

电气工程专业培养能够从事与电气工程有关的系统运行、自动控制、电力电子技术、信息处理、试验分析、研制开发、经济管理以及电子与计算机技术应用等领域工作的宽口径"复合型"高级工程技术人才。

5. 主要课程

电气工程专业的主要课程有电路原理、模拟电子技术、数字电子技术、微机原理及应用、信号与系统、自动控制原理、电机与拖动、电力电子技术、电力拖动自动控制系统、电气控制技术与PLC应用、微机控制技术和供电技术。

练习答案

1. a) It can adjust our subject direction, and improve the teaching and scientific research by learning foreign discipline state, teaching and scientific research direction.

b) Electrical automation uses relay, sensors, and electrical components to realize sequence control and time control of the process.

c) In the coming years, the biggest factors impact on the development of electrical engineering, including the following respects.

第 18 单元　数控技术简介

第二次世界大战以后，基于一些实验改变了金属的加工方法。在19世纪40年代早期，军工产品如飞机的生产需要，加快了生产技术的研究。结果，第二次世界大战结束时生产产品的形状过于复杂，难以进行实际加工。为辅助工程计算，宾夕法尼亚大学发明了ENIAC计算机，该计算机由大量的晶体管及电线组成。按照今天的标准，它难以编程并且运算速度很慢，但它已经是计算机了。然后，数字控制机床及计算机被用于开发今天的计算机数控机床。帕森斯公司，美国空军以及麻省理工学院被授权研制数控机床。1946年，帕森斯公司试图发明一种制造复杂飞机零件的方法，为了生产直升机的转子叶片，他们用具有多坐标的复杂工作台和手动机床进行实验；为了产生复杂曲线，每个坐标轴由一位操作者操控，按照顺序依次移动，这种方法效率低，容易出现错误。后来帕森斯公司致力于研究自动产生这些形状。看起来自动方法是可行的。1949年，帕森斯为空军设计了一个验证实验以证明他的想法，并说服空军签订了一个研究合同。此后不久，他与麻省理工学院合作，并在1952年，研制成功了世界上第一台三坐标数控铣床，如图18-1所示，机床的电器柜比机器还大，然而，这是加工方式改变的开始。美国麻省理工学院发明的三轴联动数控机床使用穿孔带来控制3个坐标轴运动，这台机器能迅速、准确和可靠的加工曲线外形。1960年，由于许多公司能够负担得起，制造商们开始生产数控设备，在市场上设备的价格，允许商家购买他们的第一台数控机器。

数控技术应用于工业已经50多年了。简单地说，数控技术是使用各种字母、数字和特殊字符组成的代码对机器进行自动控制的技术，为执行某个操作而编写的一套完整的编码指令被称为一个程序，程序转化为相应的电信号输入电机，驱动机器运行。用一台计算机编写

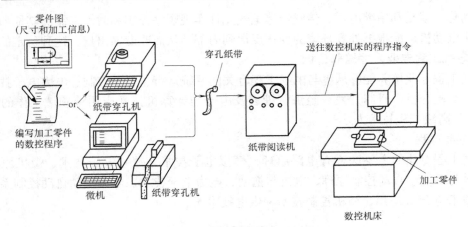

图 18-1　传统硬件数控系统的组成

程序的过程称为计算机辅助编程，本文中称为手工编程形式。传统上，数控系统，应包括以下组成部分。

纸带穿孔机：将书面指令转换成相应的孔。孔是由穿孔纸带机在纸带打的，很多老的穿孔机使用一个叫多功能打字机的设备，而新的穿孔机则使用微电脑。

磁带阅读机：将纸带上孔的信息读出，并将其转换成相应的电信号。

控制器：接收磁带阅读机输出的电信号，随后使机床回应。

处理器：响应控制器的程序信号。机器执行加工所需的动作（主轴旋转、主轴停、工作台或主轴根据程序动作等），如图 18-1 所示。

数控机床相比于普通机床有以下优点：
1. 控制刀具在最佳的切削方式下运动。
2. 提高产品的质量和可重复性。
3. 降低刀具成本、刀具磨损和作业准备时间。
4. 降低了生产时间。
5. 减少报废。
6. 有利于生产管理。

计算机数控机床（CNC）是在硬件数控机床基础上加一个内置计算机。内置计算机通常称为机器控制单元或 MCU。硬件数控机床的控制单元通常以电路相连。这意味着，机床功能由组成控制器的物理电子元件完成。另一方面，内置计算机是"集成的"。因此，在加工过程中将机床功能程序输入计算机中，即使数控机床关闭的时候也不会被删除。保存这些信息的内存被称为 ROM 或只读存储器。MCU 通常有一个键盘用于部分程序的直接或手动数据输入。这些程序存储在 RAM 中或随机存储器中。它们可以被控制器调出来，编辑和执行。存储在 RAM 中的所有程序会随着数控机床的关闭而丢失。可将这些程序存放在辅助存储设备上如穿孔带、磁带或磁盘。新 MCU 拥有图像屏幕，不仅可以显示数控程序而且可以显示刀具路径以及各种错误信息。

大多数数控系统的组成部件如图 18-2 所示。

机床控制单元：编辑、存储和处理数控程序。机床控制单元还包含了执行软件程序形式的机械运动控制器。

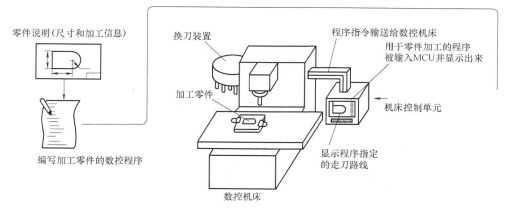

图 18-2 现代计算机数控系统的组成

数控机床：响应机床控制单元的程序信号并加工零件。

硬件数控机床必须根据程序工作，没有灵活性和可变化的自由。存储所有的信息的程序存储在外部存储器，如穿孔带上。机床控制单元的内存中存储的数据指导机床工作。机床在微处理器控制下操作，计算机控制器可执行计算与决策。

计算机数控机床比硬件数控机床具有更大的灵活性。和硬件数控机床一样，计算机数控机床根据数控程序运行。此外，它承担计算和决策的责任，包含一个存储程序的内存。这意味着应用程序可以方便地调整。可以编辑的内存称为随机存储器（RAM）。计算机数控机床能够检查错误，并帮助操作者改正错误，还可以与操作者和外部设备，如机器人、中央编程计算机进行通信。

加工中心是数控技术发展的最新产品。加工中心可以配备能够换多于 90 把刀的换刀装置。许多还配有称为托盘的矩形工位台。托盘用于自动加载和卸载工件。一个典型的加工中心可以完成铣削、钻孔、攻螺纹等功能。此外，对于更先进的加工中心可以在不同的空间、不同的角度同时完成许多任务。加工中心通过减少工作的移动来提高加工效率并节省成本，如图 18-3 所示。

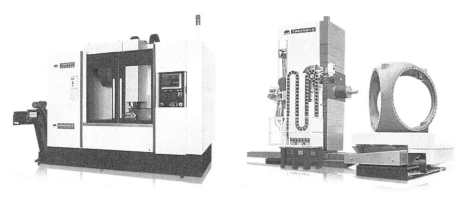

a) 立式加工中心　　　　b) 卧式加工中心

图 18-3 加工中心

车削中心由于其换刀可靠性的提高使得其有强劲的需求。这类数控机床能够执行许多不同类型的旋转车削加工操作，如图 18-4 所示。

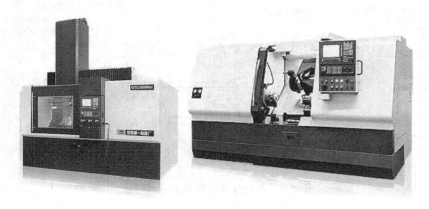

a) 立式车削中心　　　　　　　　b) 卧式车削中心

图 18-4　车削中心

除了加工中心、车削中心，数控技术也已经被应用于许多其他类型的生产设备，其中包括电火花切割机和激光切割机。电火花切割机使用一根非常细的导线作为一个电极，导线绕着两个工作辊子伸展，像带锯一样切割工件，材料与电极之间产生电火花，计算机用来控制横向工作台的动作。

激光切割数控机床采用一个强烈的光束聚焦激光切割工件。材料在激光光束作用下，温度快速上升并被蒸发，如果光束能量足够高，它将穿透材料。因为没有涉及机械切削力，激光切割工件有最少的扭曲这些机床常用于有效加工槽和钻孔。

练习答案

1. a) 总结本单元，你将能够阐述数控技术的历史。

b) 简单地说，数控技术是一种基于字母、数字和特殊字符等组成的代码来自动控制机器动作的技术。

c) 计算机数控机床是一种使用微型计算机控制的硬件数控机床。

2. a) Traditionally, numerical control systems have been composed of the following components：

tape punch, tape reader, controller, and NC machine.

b) NC systems offer some of the following advantages over manual methods of production：

- Better control of tool motions under optimum cutting conditions.
- Improved part quality and repeatability.
- Reduced tooling costs, tool wear, and job setup time.
- Reduced time to manufacture parts.
- Reduced scrap.
- Better production planning and placement of machining operations.

c) Machining centers (vertical turning center and horizontal turning center), turning centers, electrical discharge machines (wire EDM), and laser cutting machines.

阅读材料 A 单片机的节电模式

空闲节电模式

在空闲工作模式状态，CPU 保持睡眠状态而片内的其他设备仍保持激活状态，这种方式由软件产生。此时，片内 RAM 和所有特殊功能寄存器的内容保持不变。空闲模式可由任何允许的中断请求或硬件复位终止。

需要注意的是，当采用硬件复位来终止空闲工作模式时，CPU 通常是从激活空闲模式那条指令的下一条指令开始继续执行程序的，要完成内部复位操作，硬件复位脉冲要保持两个机器周期（24 个时钟周期）有效，在这种情况下，禁止 CPU 访问片内 RAM，而允许访问其他端口。为了避免在复位结束时可能对端口意外写入，激活空闲模式的那条指令的后一条指令不应是一条对端口或外部存储器的写入指令。

掉电模式

在掉电模式下，振荡器停止工作，进入掉电模式的指令是最后一条被执行的指令，片内 RAM 和特殊功能寄存器的内容在终止掉电模式前被冻结。退出掉电模式的方法是硬件复位或由处于使能状态的外部中断 INT0 和 INT1 激活。复位后将重新定义全部特殊功能寄存器但不改变片内 RAM 中的内容，在 VCC 恢复到正常工作电平前，复位无效，且必须保持一定时间以使振荡器重启动并稳定工作。

阅读材料 B 设计一个微型 PLC 系统的指导原则

设计一个微型 PLC 系统有许多设计方法。以下这些通用的指导原则适用于许多设计项目。当然，您所在公司的规程和您在培训中接受的实践经验是必须遵循的。

分解控制过程或者机器

将您的控制过程或者机器分解成相互独立的部分。分解决定了控制器之间的界限，并将影响功能描述和资源的分配。

创建功能说明

写出过程或者机器每一部分的操作描述。它包括以下内容：I/O 点、操作的功能描述、每个执行机构（例如线圈、电动机和驱动器等）在动作之前需要满足的状态、操作接口的描述以及过程或机器其他部分的接口描述。

安全电路的设计

识别要求设计硬件安全线路的设备。控制设备在不安全的条件下出现故障，会造成不可预料的启动或者机器操作的变化。在不可预料或者不正确的机器操作会造成人身伤害或严重财产损失的场合，应该考虑采用和 S7-200 独立的机电冗余来防止不安全的操作。

指定操作员站

根据功能描述的要求建立操作员站的配置图。

创建配置图

根据功能描述的要求建立控制设备的配置图。包括如下内容：
- 和过程或者机器有关的每个 S7-200 的位置图。
- S7-200 和扩展 I/O 模块的机械布局图（包括控制柜和其他设备）。
- 每个 S7-200 和扩展模块的电气图（包括设备型号、通信地址和 I/O 地址）。

建立符号名表（可选）

如果选择了符号名寻址，需要对绝对地址建立一个符号名表。符号名表不仅包括物理输入/输出信号，还包括程序中用到的其他元件。

阅读材料 C 液压传动

对于两点之间较远的传动，不适合用传动带和传动链传动的机械系统，可优先考虑采用液压传动，液压传动的优点：低速大转矩，结构紧密，稳定性好，滑动平稳，速度和方向控制灵活，可实现无级化输出。

由电力驱动的油泵，提供有传递能量作用的压力油，并可以供给液压马达或油缸，从而将液压能转化为机械能。液压油流动是通过控制阀进行控制的，压力油流动产生线性的或者螺旋性的机械运动，此时油液的动能较低。因此，有时候使用静压传动。液压马达与液压泵的结构几乎是一样的，任何液压泵都可以当做马达使用，一定时间内油的流量可通过调节阀门或使用变量泵来控制。一般来说，静压传动可以分为直线和螺旋的，螺旋式传动产生螺旋运动，而由活塞及缸体部件产生往复的运动是线性运动。

所有液压马达的功能基本源于一个原理：压力油被交替地压入、压出油腔中，进油循环由最小的腔体注油开始，当腔体达到最大容积时，油腔和油路隔开，停止进油；压力油通过回油管返回到油箱中，同时另一个油腔开始进油。

阅读材料 D 数控机床用刀具

计算机数控机床最令人吃惊的是快速切削技术的使用，切削刀具能抵抗这样的切削力看起来很不可思议，另外数控机床检索时间少于 1s，换刀时间大约为 5s，这就很容易理解为什么许多产品工程师将刀具作为计算机数控机床的最重要的影响因素。

虽然高速钢（HSS）常用于小直径钻头、丝锥、铰刀、平铣刀和中心钻等，但是现在大部分计算机数控机床刀具使用硬质合金。

切削工具必要的物理性质包括：好的红硬性（即使高达 600℃ 的高温）和高的韧性。高速钢与硬质合金相比，韧性好但硬度不够，不适宜高速切削。然而另一方面，由于硬质合金缺乏韧性，这就意味着需要开展大量的研究，提高硬质合金的硬度等级，以满足现代加工技术的要求。

硬质合金的硬度几乎相当于钻石的硬度，硬度源自于其结构—碳化钨。纯碳由于易碎不宜用于切削刀具，所以需要添加微量元素，将碳化物和钴做成需要的形状，然后烧结熔化，并与碳化物凝固成稠状、无孔的结构。

除了碳化钨，其他硬质材料，如钛、钽合金也被使用，在硬质合金工具上涂有一层薄薄的碳化钛，工具的耐磨性和使用寿命将大大增加，可达到原来的五倍。

数控机床为满足生产，使用各种刀具，这些刀具被认为是一个刀具系统，如何快速找到并将刀具安全地固定在某个位置被看做是机床刀具制造商的一个重要指标。

加工中心是这样一个系统，它带有自动换刀装置，可以方便快速地实现换刀动作。

自动换刀系统的特点：

1. 换刀方便快捷。
2. 确保刀具能重复使用。

3. 提供快速容易退刀的路线。

我们已经设计了许多不同类型的存储和换刀装置，其中最具代表性的三种是：转塔式、圆盘式和链式。换刀系统有立式和卧式的。

第一种类型的主要用于老式数控钻床。刀具通常存储在刀库中，当程序调用时，刀具转到指定位置，快速地将刀具装入主轴。虽然这种设计实现了快速换刀，但是限制了刀库的容量。

第二类系统通常在立式加工中心。刀具被存储在一个称为轮鼓的码盘的圆盘中，圆盘在空间旋转至当前刀具的位置，卸下当前刀具，继续旋转，将新刀具送到指定位置并装入主轴，对于更大的系统，主轴移动到圆盘换刀。

自 1972 年以来链式刀库已广泛应用于加工中心。这种类型的刀库允许一个比较小空间里有许多刀具。链条可能位于数控机床的旁边或顶端。这样使刀具远离主轴和工件，可以保证最少的换刀时间并最大限度地保护刀具。假如输入一个换刀程序，系统在链中可以准确地找到相应的刀具，旋转臂将刀具从刀库中取出，并将原来的刀具从主轴上卸下，然后在刀库中为卸下的刀具找一个位置，旋转臂再次旋转并且将新刀装入主轴中，老刀具入库，最后旋转回位。

目前有两种方法识别使用的刀具。一个是采用条形码的方式，这种刀具识别方法是当程序需要一个特定的刀具时，控制器寻找一个特定的刀具代码，而不是一个特定的位置。另一种刀具识别系统，采用了计算机芯片，事先将刀具识别数和一部分刀具的相关信息通过微芯片存储起来，再由一个特殊的传感器读出相应的信息送到控制器。

第 V 部分　工程应用

第 19 单元　WinCC flexible 介绍

1. SIMATIC HMI 介绍

简介

在工艺过程日趋复杂、对机器和设备功能的要求不断增加的环境中，获得最大的透明性对操作员来说至关重要。人机界面（HMI）提供了这种透明性。

HMI 是人（操作员）与过程（机器/设备）之间的接口。PLC 是控制过程的实际单元。因此，在操作员和 WinCC flexible（位于 HMI 设备端）之间以及 WinCC flexible 和 PLC 之间均存在一个接口。HMI 系统承担下列任务。

- 过程可视化

过程显示在 HMI 设备上。HMI 设备上的画面可根据过程变化动态更新。

- 操作员对过程的控制

操作员可以通过 GUI（图形用户界面）来控制过程。例如，操作员可以预置控件的参考数值或者启动电机。

- 显示报警

过程的临界状态会自动触发报警，例如，当超出设定值时。

- 归档过程值和报警

HMI 系统可以记录报警和过程值。该功能使您可以记录过程序列，并检索以前的生产数据。
- 过程值和报警记录

HMI 系统可以输出报警和过程值报表。例如，您可以在某一轮班结束时打印输出生产数据。

- 过程和设备的参数管理

HMI 系统可以将过程和设备的参数存储在配方中。例如，可以一次性将这些参数从 HMI 设备下载到 PLC，以便改变产品版本进行生产。

SIMATIC HMI

SIMATIC HMI 提供了一个全集成的单源系统，用于各种形式的操作员监控任务。使用 SIMATIC HMI，您可以始终控制过程并使机器和设备持续运行。

用于设备级的小型触摸面板是简单 SIMATIC HMI 系统的应用实例。

用于监控生产工厂的 SIMATIC HMI 系统代表了拥有高端性能范围的产品。它们包括高性能的客户机/服务器系统。

SIMATIC WinCC flexible 的集成

WinCC flexible 是一种前瞻性的面向机器的自动化概念的 HMI 软件，它具有舒适而高效的设计。WinCC flexible 综合了下列优点：

- 直接的处理方式
- 透明性
- 灵活性

2. WinCC flexible 系统概述

（1）WinCC flexible 的组件

WinCC flexible 工程系统

WinCC flexible 工程系统是用于处理所有基本组态任务的软件。WinCC flexible 版本决定了在 SIMATIC HMI 系列中可以组态哪些 HMI 设备。

WinCC flexible 运行系统

WinCC flexible 运行系统是用于过程可视化的软件。在运行系统中，您可以在过程模式下执行项目。

WinCC flexible 选件

WinCC flexible 选件可以扩展 WinCC flexible 的标准功能。每个选件需要一个单独的许可证。

（2）WinCC flexible 工程系统

简介

WinCC flexible 是用于所有组态任务的工程系统。WinCC flexible 采用模块化的设计。随着版本的逐步升高，所支持的设备范围以及 WinCC flexible 的功能都得到了扩展。您也可以通过 PowerPack 程序包将项目移植到更高版本中。

WinCC flexible 包括了性能从 Micro Panel 到简单的 PC 可视化的一系列产品。因此，WinCC flexible 的功能可以与 ProTool 系列的产品和 TP Designer 相媲美。您可以将现有的 ProTool 项目集成到 WinCC flexible 中。

练习答案

1. 过程可视化；设定值；直接的处理方式；灵活性；从……变化到……
2. Process visualization; Operator control of the process; Displaying alarms; Archiving process values and alarms; Process values and alarms logging; Process and machine parameter management.
3. a) 通过过程总线直接与PLC连接的HMI设备称为单用户系统。单用户系统通常用于生产，但也可以配置为操作和监视独立的部分过程或系统区域。

   b) 带多台HMI设备的PLC：多台HMI设备通过过程总线（例如PROFIBUS或以太网）连接至一个或多个PLC。例如，在生产线中配置此类系统以从多个点操作设备。

   c) 具有集中功能的HMI系统：HMI系统通过以太网连接至PC。上位PC机承担中心功能，例如配方管理。必要的配方数据记录由次级HMI系统提供。

第20单元　柔性制造系统

柔性制造系统把制造中的所有主要元件集合成为一个高自动化的系统。首次应用于20世纪60年代末期，它由一系列的制造单元组成，每个单元包含一个工业机器人（服务于多个CNC系统）和一个自动物料处理系统，这些都由一台中央计算机控制。制造过程中的不同计算机指令可以下载并通过工作站依次传输给零件。

整个系统自动化程度很高，它能优化整个制造过程中的每一步。这些步骤可能包括一个或多个程序和操作（比如加工、磨削、切削、成型、粉末冶金、热处理和修整），还有原材料的处理、检查和装配。到目前为止，FMS最常见的应用就是在加工和装配操作中，从机床制造中可以获得许多的FMS技术。

在制造行业中，柔性制造系统代表着高效、高精度和高生产力，FMS的柔性体现在它能处理各种不同的外形轮廓和以任意的顺序加工。

FMS可以认为包含了另外其他两个系统的优点：（a）高生产率但固定传输的生产线；（b）可以在独立的机器上生产大量多样化产品但生产效率低的加工生产车间。

FMS单元

一个柔性制造系统的基本元素是：（a）工作台；（b）原材料和部件的自动处理和输送；（c）控制系统。通过这个系统，物料、部件、产品进行有序流动，在生产过程中工作台的工作效率最大。

车间里的机器类型取决于产品的类型。对于机械加工操作，它们由多个三轴或五轴加工中心、CNC车削、铣削加工、钻孔和磨削组成。同样还包括其他各类设备，比如自动检测、装配和清洗。

其他适合FMS的操作包括板材成型、冲压和裁剪、锻造。它们把熔炉、锻造机器、裁剪、热处理设备和清洗设备融于一体。

由于柔性制造系统的柔性，物料处理、存储和回收等系统显得很重要。物料处理由一台中央计算机控制并由自动引导车、传送器和不同的传送机制执行。这个系统能够传输原材料、数据块，以及零件在任一机床任何时间不同阶段的完成情况。棱柱型部件通常在专门设计的托盘上传送。具有旋转对称的部件（比如用于车削加工的部件）通常由机械装置和机器人来传送。

　　FMS 的计算机控制系统相当于它的大脑，包括许多硬件和软件，子系统控制车间的机器设备和原材料、数据块、零件等在加工的不同阶段从一台机器到另一台机器的传送。它同样储存数据和提供可视化数据的通信终端。

安排调度

　　因为 FMS 涉及巨大的资金投资，高效的机器利用率是问题的本质：机器绝对不能闲置。接下来，适当的计划和过程安排很重要。

　　FMS 的计划安排是动态的。不像一般的车间，在那里遵循一个相对刚性的安排去执行一系列的操作，FMS 的日程安排计划系统具体说明了用哪种类型的操作去执行每一个零部件。它指定了需要用到的机器和制造单元。动态的计划安排能够对产品类型的迅速变化作出反应，并且对实时决策反应也是灵敏的。

　　因为 FMS 的柔性，在制造过程中的转化可以省去安装时间，这个系统能够在不同的机器以不同的顺序执行不同的操作。但是，必须对系统中的每个制造单元的特点、性能和可靠性进行检验，以确保在工作站之间流动的工件满足质量和尺寸精度方面的要求。

FMS 在经济上的合理性

　　FMS 装置都是资金密集型的，一般起价都超过一百万美元，那么，在作决定之前一个全面的代价—回报分析必须执行。这个分析应该包括如下因素：资金、能源、材料和劳动力的投入，即将生产的产品的预期市场，以及市场需求和产品种类的预期浮动。另外一个因素是花费在安装系统和排除系统故障上面的时间和精力。

　　一般地，一个 FMS 系统需要 2～5 年的安装时间和至少 6 个月的调试时间，尽管 FMS 需要很少的机器操作，但是负责整个操作的工作人员必须经过培训并有很熟练的技术。这些人员包括制造工程师、计算机程序员和维修工程师。

　　与常规的制造系统相比，FMS 的优点如下：
　　（1）零件可以随意加工，不管是批量还是单件生产，并且花费少；
　　（2）直接劳工成本和库存减少，比起常规系统节约了许多；
　　（3）产品变更所需的交货时间缩短了；
　　（4）产品更加可靠，因为系统有自动纠错能力，而且产品质量都是一致的。

练习答案

　　1. 制造单元；工业机器人；自动物料处理系统；生产车间；资金密集型。

　　2. The basic elements of a flexible manufacturing system are (a) workstations, (b) automated handling and transport materials and parts, and (c) control system. The workstations are arranged to yield the greatest efficiency in production, with an orderly flow of materials, parts, and products through the system.

　　3. a) FMS 基本上是一个由自动化设备、传送带、计算机组成的加工车间。这种系统的调度非常复杂，由于不同零件的加工时间相差很大，很难将 FMS 与一个集成系统相连接，使得 FMS 通常成为花费昂贵的自动化孤岛。

　　b) FMS 的发展始于 20 世纪 60 年代的美国，其设计思想是把生产线的高可靠性和高生产率同数控机床可编程的灵活性相结合，以便能生产各种零件。

c) 一个 FMS 系统每班需要三个或四个工人来装卸零件、更换刀具和进行日常性维护，FMS 系统的工人通常都是经过数控和计算机数控培训的高级技术人员。

第 21 单元　CAD/CAM/CAPP

　　CAD 是 Computer Aided Design 的缩写，中文意思是计算机辅助设计，指以计算机为辅助手段来完成整个产品的设计过程。广义的 CAD 包括设计和分析两个方面。典型的 CAD 硬件包括计算机、一个或更多的图形显示终端、键盘以及其他外围设备。CAD 软件则由完成系统的计算机图形处理的程序加上一些使用户公司的工程功能变得容易的应用程序组成。这些应用程序的例子包括零件应力-应变分析，动态反应机制，热传递计算，以及数控零件编程。从一个用户公司到下一个用户公司的应用程序的集合可能会有所不同，因为他们的生产线、制造工艺和客户市场是不同的。这些因素都增加了 CAD 系统要求的不同。

　　CAM 是 Computer Aided Manufacturing 的缩写，中文意思是计算机辅助制造，广义的 CAM 是指通过计算机与生产设备直接的或间接的联系，进行规划、设计、管理和控制产品的生产制造过程。主要包括使用计算机来完成数控编程、加工过程仿真、数控加工、质量检验、产品装配和调试等工作。CAM 的核心是计算机数值控制（简称数控），是将计算机应用于生产制造的过程或系统。1952 年美国麻省理工学院首先研制成数控铣床。数控的特征是由编码在穿孔纸带上的程序指令来控制机床。此后发展了一系列的数控机床，包括称为"加工中心"的多功能机床，能从刀库中自动换刀和自动转换工作位置，能连续完成铣、钻、铰、攻螺纹等多道工序，这些都是通过程序指令控制运作的。只要改变程序指令就可改变加工过程，数控的这种加工灵活性被称为"柔性"。

　　CAPP 是 Computer Aided Process Planning 的缩写，中文意思是计算机辅助工艺规划，是指借助于计算机软硬件技术和支撑环境，利用计算机进行数值计算、逻辑判断和推理等功能，来制定零件机械加工工艺过程。操作者通过向计算机输入被加工零件的原始数据、加工条件和加工要求，由计算机自动地进行编码、编程，直至最后经过优化的工艺规程卡片输出。这项工作需要有丰富生产经验的工程师进行复杂的规划，并借助计算机图形学、数据库以及专家系统等计算机科学技术来实现的。计算机辅助工艺规划常是联结计算机辅助设计（CAD）和计算机辅助制造（CAM）的桥梁。

　　CAPP 技术的研究和发展源于 20 世纪 60 年代。1969 年挪威推出了世界上第一个 CAPP 系统 AUTOPROS，并于 1973 年商品化。美国于 20 世纪 60 年代末 70 年代初着手于 CAPP 系统。1976 年美国的国际计算机辅助制造公司 CAM-I（Computer-Aided Manufacturing-International, Inc）所推出的 CAPP 系统最著名、应用最广泛，在发展历史上具有里程碑意义。此后，世界上有众多 CAPP 系统问世，上海同济大学在 1982 年开发出我国第一个 CAPP 系统，即 TOJICAP 系统。

　　CAPP 系统由五大模块组成：零件信息的获取、工艺决策、工艺数据库/知识库、人机界面和工艺文件管理/输出。

　　CAPP 系统按其工作原理可分为检索式、派生式和创成式三种。检索式 CAPP 系统适用于标准工艺，你可以事先对设计好的零件标准工艺进行编号，然后将它预存在计算机中，当你想制定零件的工艺过程时，可根据输入的零件信息进行搜索，查找合适的标准工艺。根据"相似的零件有相似的工艺过程"这一原理，派生式 CAPP 通过检索相似典型零件的工艺过程，加以增删或编辑而派生一个新零件的工艺过程。创成式 CAPP 系统和派生式 CAPP 系统不同，它根据输入的零件信息，依靠系统中的工程数据和决策方法自动生成零件的工艺过程。

CAPP 的优点在于：①可以将工艺设计人员从繁琐和重复性的劳动中解脱出来，以更多的时间和精力从事更具创造性的工作。②可以大大缩短生产准备周期和工艺设计周期，提高企业对瞬息变化的市场需求作出快速反应的能力，提高企业产品在市场上的竞争能力。③有助于对工艺设计人员的宝贵经验进行总结和继承。④有利于对工艺设计的最优化和标准化，提高工艺设计质量，提高生产率，减少制造成本。⑤为实现企业信息集成创造条件，进而便于实现并行工程、敏捷制造等先进生产制作模式。

借助于 CAPP 系统，可以解决手工工艺设计效率低、一致性差、质量不稳定、不易达到优化等问题，无论是对单件小批多品种生产还是对大批量生产都有重要意义。

纵观 CAPP 发展的历程，可以看到 CAPP 的研究和应用始终围绕着两方面的需要而展开：一是不断完善自身在应用中出现的不足；二是不断满足新的技术、制造模式对其提出的新的要求。因此，未来 CAPP 的发展，将在应用范围、应用的深度和水平等方面进行拓展，具体表现为以下三方面的发展趋势：一是面向产品全生命周期的 CAPP 系统；二是基于知识的 CAPP 系统；三是基于平台技术、可重构式的 CAPP 系统。

练习答案

1. a）CAD、CAM、CAPP 是否都需要计算机来辅助完成？是。

b）CAPP 常在 CAD 和 CAM 之间充当桥梁角色？是。

c）哪个国家率先推出了世界上第一个 CAPP 系统？挪威。

d）请描述一下 CAPP 的优点。CAPP 的优点在于：①可以将工艺设计人员从烦琐和重复性的劳动中解脱出来，以更多的时间和精力从事更具创造性的工作。②可以大大缩短生产准备周期和工艺设计周期，提高企业对瞬息变化的市场需求作出快速反应的能力，提高企业产品在市场上的竞争能力。③有助于对工艺设计人员的宝贵经验进行总结和继承。④有利于对工艺设计的最优化和标准化，提高工艺设计质量，提高生产率，减少制造成本。⑤为实现企业信息集成创造条件，进而便于实现并行工程、敏捷制造等先进生产制作模式。

2. 设计　　间接的　　仿真　　产品装配　　判断　　逻辑
加工工艺　　编程　　数据库　　模块　　标准化　　大批量生产

3. a) By means of a CAPP system, we can solve the low efficiency of manual process design, poor consistency, unstable quality, and not easy to achieve optimization.

b) Shanghai Tongji University developed the first CAPP system of China in 1982.

第22单元　IRB 140 工业机器人数据手册

IRB 140 是 ABB 公司的一款工业机器人，它的主要应用领域包括弧焊、装配、清理/喷雾、上下料、物料搬运、包装和去毛刺等。

小巧，强劲，快速

外形紧凑、功能强劲的 IRB 140 是一款六轴多用途工业机器人，有效荷重 6kg，工作范围长达 810mm，可选落地安装、倒置安装或任意角度挂壁安装方式，见表 22-1。IRB 140 分标准型、铸造专家型、洁净室型、可冲洗型 4 种机型,所有机械臂均全面达到 IP67 防护等级,易于同各类应用相集成与融合。IRB 140 上臂采用后翻转机构,即使采用挂壁安装,第 1 轴仍可旋转 360°,工作半径显著扩大。

表 22-1　IRB 140 系列机器人规格和性能参数

规格			
机器人型号	荷重能力	第 5 轴工作范围	注释
IRB 140/IRB 140T	6kg	810mm	
IRB 140F/IRB 140TF	6kg	810mm	铸造专家型防护
IRB 140CR/IRB 140TCR	6kg	810mm	洁净室型
IRB 140W/IRB 140TW	6kg	810mm	可冲洗型
附加荷重(上臂或手腕)			
上臂		1kg	
手腕		0.5kg	
轴数			
机器人本体		6	
外部设备		6	
集成信号源	上臂 12 路信号		
集成气源	上臂最高 8bar(气压单位)		
性能			
重复定位精度	0.03mm(ISO 实验平均值)		
轴动作	轴	工作范围	
	1.(C 旋转)	360°	
	2.(B 手臂)	200°	
	3.(A 手臂)	280°	
	4.(D 手腕)	无限制(默认 400°)	
	5.(E 弯曲)	240°	
	6.(P 翻转)	无限制(默认 800°)	
TCP 最大速度(工具端中心)		2.5m/s	
TCP 最大加速度		20m/s^2	
加速时间 0~1m/s		0.15s	
速度			
轴号	**IRB 140**	**IRB 140T**	
1	200°/s	250°/s	
2	200°/s	250°/s	
3	260°/s	260°/s	
4	360°/s	360°/s	
5	360°/s	360°/s	
6	450°/s	450°/s	

（续）

节拍时间		
5kg 拾料侧	IRB 140	IRB 140T
节拍 25×300×25mm³	0.85s	0.77s
电气连接		
电源电压	200~600V, 50/60Hz	
额定功率		
变压器额定值	4.5kV·A	
典型功耗	0.4 kW	
物理特性		
机器人安装	任意角度	
尺寸		
机器人底座	400×450mm²	
机器人控制器（高×宽×深）	950×800×620mm³	
重量		
机器人本体	98kg	
环境		
机器人本体环境温度	5~45°C	
相对湿度	Max. 95%	
机器人本体防护等级	IP67	
选件	铸造件	
	可冲洗型（耐高压蒸汽清洗）	
	洁净室型，6级（IPA认证）	
噪声水平	最高 70dB（A）	
安全	带监控、紧急停和安全功能的双回路	
	3位启动装置	
辐射	EMC/EMI 屏蔽	

数据和尺寸若有变更，恕不另行通知。

工作范围

工作范围如图 22-1 所示。

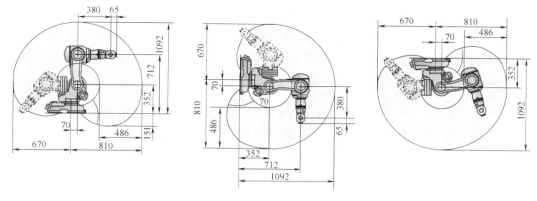

图 22-1 IRB 140 机器人工作范围

IRB 140 设计紧凑、牢靠，采用集成式线缆包，进一步提高了整体柔性。可选配碰撞检测功能（实现全路径回退），使可靠性和安全性更有保障。

IRB 140T 以第 1、第 2 轴作业为主，大幅缩短节拍时间。在使用第 1、第 2 轴的场合下，节拍时间可缩短 15%～20%。这款高速型产品配套 PickMaster，是包装作业和引导式作业的理想之选。

IRB 铸造专家型和可冲洗型适合在极端铸造环境及其他对抗腐蚀性和密封性要求严苛的恶劣环境中使用。两种机型均达到 IP67 防护等级，并经高标准表面处理，耐高压蒸汽清洗。白色涂装的洁净室型机器人则达到 10 级洁净室标准，尤其适合对洁净度有严格要求的生产环境。

练习答案

1. 落地安装；铸造专家型；洁净室型和可冲洗型；碰撞检测；抗腐蚀性和密封性；EMC/EMI 屏蔽

2. a) 目前机器人的研究主要在 2 个领域：人工智能和机器视觉相关的领域。机器人技术中一个典型的人工智能问题可能是机器人在一个杂乱的环境中找到一个明确的路径。然而，这一领域大多数研究是制造智能机器。这是单纯的人工智能，和机器人技术没有关系，除非立即成功应用于机器人。

b) 让机器人动作更快就有必要使连杆更轻。这意味着它们将更加灵活。在本书中我们假定连杆是完全刚性的，当然这是近似的。机器人的许多工作具有显著的灵活性。这对外太空机器人特别重要，减少重量显得非常珍贵。减少弹力引起的振动对于地球上精确的工作也至关重要。

第 23 单元　机器视觉

导言

机器视觉（MV）是融合了机械、光学、电子和软件的系统工程。它可应用于原材料、工业设备和生产过程的检验以及产品缺陷的探测，以提高产品质量和生产效率、保障产品和生产过程的安全，还可应用于生产过程的监控。

MV 系统根据功能可以分成三类：视觉识别系统（VRS）、视觉检测系统（VIS）和视觉导航系统（VGS）。在实际应用中系统可以是两类或三类的组合，如 VRS + VIS、VRS +

VGS、VGS + VIS 或者 VRS + VIS + VGS。

VRS 强调的是在给定的区域内识别目标。典型的应用是车辆识别系统，残疾人通过具有视觉功能的机器人进行目标识别，VRS 还应用在医学、艺术、环境监测和导航领域中。

VIS 通常用于检测，如测定产品的变形、误差和缺损等。这类系统强调的是检测目标的尺寸和形状，能 100% 的实时检验和测量是十分重要的。在工业领域中有许多成功的应用实例，如气门杆油封视觉检测系统、三维机械零件视觉检测系统和自动 PCB 检测等。

VGS 强调的是目标定位和引导控制器移动物体。这类视觉系统的测量回路，提供目标的位置和方向。自动化程度较高的企业都有应用 VGS 的潜力，如路虎汽车的原料供料和汽车筛管下入系统。

MV 系统的原型如图 23-1 所示，可看出机器视觉融合了多门学科技术。

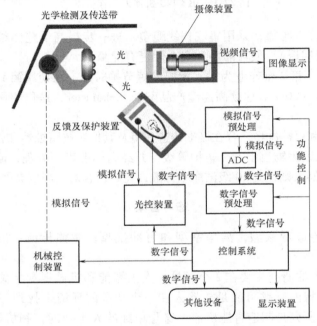

图 23-1　MV 原型系统

机器视觉系统要素

机器视觉必然涉及了如下领域方面的研究：机械搬运、照明、光学、图像传感器、电子学、信号处理、图像处理、数字系统结构、软件、工业工程、人机接口、控制系统和制造业。

机器视觉在工业中的应用

机器视觉在工业中的应用主要有检测、机器人导航、过程监测和控制，我们将在下面分别讨论。

检测

检测应用在两个方面，狭义上指的是检测缺陷，如原料的缺陷、加工的不完善（如没开孔、斜面不够、抛光不完整、螺钉没开槽等）、误标记等，广义上视觉检测包括计算、分级、归类、定位和识别等。

自动视觉检测（AVI）和机器人视觉间没有明显的区别，因为许多检测任务需要操作零件，而许多目标操作应用也需要识别和验证。

机器人视觉

"机器人"是指那些在计算机的控制下，能够移动末端执行器（比如夹具、摄像机等）到给定的二维或三维空间中的点的机械。机器人也可以控制末端执行器的方位。根据这个定义，下面的机械装置可被认定为机器人，如数控磨床、钻床、车床、电子元件插入机、绘图机、蛇形机器人手臂、自动导航车等。

最初的工业机器人被称为"盲奴隶"，它完成重复性的任务，能有效处理工件的外形和位置可预测的情况，然而对处理新的、不可预料的情况是无能为力的。机器人的位置精度很高，它能够在不知道工件位置和姿态的情况下、在工件尺寸和外形多样性的情况下完成具体的操作。视觉就理所当然成为理想的定位部件，指导机器人完成任务。如果没有正确感知周围的环境，机器人可能处于危险状态，它可能损坏自己或其他装置、部件甚至伤害周围的人，很明显视觉是主要的感知方法。视觉引导机器人可以看到前面并且在保证没有危险的操作后决定是否安全移动。举个例子，一个智能视觉引导机器人，能够捡起随意丢在桌子上的文件，尽管它之前没有看过这文件，如果没有视觉引导，这是不可能的。

过程监测和视觉控制

视觉反馈/前馈为那些不能使用其他传感器技术的过程控制提供了机会，如通过安放摄像机检测产品（半成品或成品）、制造过程甚至废品。

图23-2中，摄像机用于监测生产，值得注意的是除了接受/拒绝控制线外，从图像处理子系统还产生了两路输出信号。

反馈是指基于上游机械调整制造过程的操作参数。

前馈是指基于下游机械调整制造过程的操作参数。

在这种情况下，摄像机被安装在与生产过程相隔的空间，使它可检测部分完工的产品，这类产品既可以是分离的单件产品，也可以是连续的多件产品。

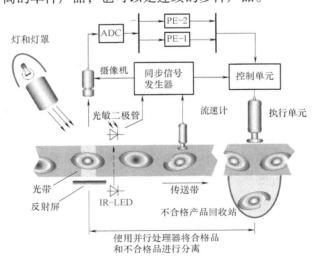

图23-2 线路控制

结论

机器视觉系统正在被广泛应用于机械工业生产中。如今，电子工业已经成为最热衷应用机器视觉的领域，而在汽车、玻璃、塑料、飞机、印刷、制药、食品、医疗产品等行业中也有了越来越多的应用，机器视觉技术本身已经显示出它能够提供更广泛的应用。

练习答案

1. a) Visual recognition system (VRS), visual inspection system (VIS), and visual guidance system (VGS).

b) Typical applications of the VRS are vehicle recognition systems, object recognition by robots with vision function for the disabled and in the fields of medicine, arts, environmental monitoring and aviation.

VIS is commonly used in inspection applications for detecting deformations, deviations from specifications, or missing elements. For example, the 100 percent visual inspection system of valve-stem seals, 3D mechanical parts visual inspection and automatic PCB inspection.

c) Light, Lighting controller, Mechanical sub-system controller, Image display, Intelligent image processing plus system control and so on.

2. 视觉识别系统　视觉检测系统　视觉导航系统　图像传感器　信号处理　工业工程　缺陷　自动视觉检测　二维　重复的　优先于　智能视觉导航机器人

3. a) MV 系统根据功能可以分成三类：视觉识别系统（VRS）、视觉检测系统（VIS）和视觉导航系统（VGS）。

b) "检测"这个词被用于两个方面：狭义上指的是检测缺陷，如原料的缺陷、加工的不完善（如没开孔、斜面不够、抛光不完整、螺钉没开槽等）、误标记等，广义上视觉检测包括计算、分级、归类、定位和识别等。

c) 视觉反馈/前馈为那些不能使用其他传感器技术的过程控制提供了机会，安放摄像头以便观测产品（半成品或成品）、制造过程甚至废品。

d) 如今，电子工业已经成为最热衷应用机器视觉的领域，而在汽车、玻璃、塑料、飞机、印刷、制药、食品、医疗产品等行业中也有了越来越多的应用。

第 24 单元　自动装配线

产品概述

人们对商品日益增加的需求导致工程师们要不断探究和开发新的生产方法，许多制造技术分支的个体发展已经能够允许以较低的成本投入来提高产量。装配工艺是制造工艺中最重要的一个环节，这个过程主要是将两个或者更多的单元部件有机组合在一起制成产品。

装配工艺发展的早期历史是和大生产发展的历史密切相关的，因而大生产的先驱也是现代装配工艺的先驱，他们新的创意和理念对大批量生产过程中采用的装配方法产生巨大的影响。

然而，尽管许多制造工程分支如金属切削和金属成形工艺已经得到迅猛发展，但是以基础的装配工艺技术却落后了很多。表 24-1 显示出在美国，装配工艺中所使用的劳动力占总劳动力的百分比从农业机械制造业中的约 20% 到电话和电信设备制造业中的几乎 60% 的变化。因此，装备工艺成本占到制造成本的 50% 以上，统计调查显示这个百分比数字仍在逐年上涨。

表 24-1　装配工艺中所使用的劳动力占总劳动力的百分比

工业	装配工艺中使用劳动力所占百分比
机动车辆	45.6
飞机	25.6
电话电报	58.9
农业机械	20.1
家用冰箱和冰柜	32.0
打字机	35.9
家用烹饪设备	38.1
摩托车自行车及配件	26.3

在过去的几年里，技术人员通过应用自动化和现代技术（如超声波焊接和压铸）不断努力地减少装配工艺的成本。然而这些功效是有限的，原因是很多装配操作者仍然使用和工业革命时期相同的工具。

早期制造技术中，产品的全部组装仅由一位操作者完成，通常这位操作者也制造装配零件。所以，这就要求操作者必须是这项工作全方面的专家，培训一个新的操作者需要花费大量的时间和费用。生产规模不是受限于产品的需求，而是常常受限于可使用的训练有素的操作者的人数。

在 1798 年，美国需要大量枪支但是联邦军工厂无法满足这种需求，由于与法国的战争迫在眉睫，不能从欧洲获取额外的枪械供应。然而，艾利惠特尼，现在已经被认为是大生产的先驱，承诺可以在 28 个月内提供 1 万只火枪，尽管他花了十年才能完成这个合同。惠特尼的新奇创意在大生产中被证明是成功的，位于康涅狄格州纽黑文的工厂，专门为制造火枪而建，有生产可互换零件的设备，这些设备不仅减少了对各类操作者技能的要求也大大提高了生产效率。在 1801 年历史证明，当惠特尼任意地从一堆元件中选取部件组装成滑膛枪锁时，让他的贵宾大吃一惊。

艾利惠特尼的工作给制造方法带来了三个主要进步。第一，机器制造出的零件比手工制作的零件具有稳定的更高的质量。这些零件具有可互换性，因此简化了装配工作。第二，最终产品的精度可以保持在一个较高的标准。第三，生产率可以大大提高。

奥利弗埃文斯的不用手工从一个地方到另一个地方传送原料的理念最终使得自动装配技术进一步发展。1793 年，他在自动面粉磨机上使用三种类型传送机仅需要两个操作工。一个操作工将谷物倒进料斗，另一个操作工对磨好的面粉进行满带包装，所有的中间作业都由传送机携带原料从一个操作到另一个操作自动地进行。

下一个对自动装配技术的发展作出杰出贡献的是伊莱休鲁特。1849 年，伊莱休鲁特加入了一家生产六发式科尔特手枪的公司。尽管那时组装单元部件的各种操作都很简单，他把这些操作划分成独立的基本单元，这些单元更容易被正确快速完成。鲁特的分类操作法提出了一种细化工作、倍增输出的理念。利用这个原则，装配工作被简化成最基本的操作，只需要经过简单培训的操作工就可胜任，而且工作效率大大提高。

弗雷德里克温斯洛泰勒可能是第一位将时间和动作研究法引入制造技术的人，其目的是通过保证相关的工件放置在易于执行所需任务的位置来节省操作工的时间和精力，泰勒也发现任何工序都有最佳的工作速度，如果超时了，会导致整体性能的下降。

无疑，对生产和装配技术发展最主要的贡献者是亨利福特。他描述的装配原则是：首先，放置工具然后将人按操作工艺排序，使得在整个工艺过程中每个工件的传递距离最短；其次，使用工件滑道或其他形式的输送装置，以使得当一个工人完成其操作时能一直在其最顺手的同一位置放下零件，并且如果可能的话，让重力将零件送至下一个工人。最后，使用滑道装配线，被组装的部件以恰当的时间间隔被传送，间隔使得工作更加容易。

这些原则后来逐渐被应用在T型福特汽车生产上。

练习答案

1. a) Percentages of producing workers involved in industry assembly, such as motor vehicles, aircraft, telephone and telegraph, farm machinery, household refrigerators and freezers, typewriters, household cooking equipment, motorcycles bicycles and parts.

   b) Eli Whitney.

   c) Henry Ford.

   d) First, place the tools and then men in the sequence of the operations so that each part shall travel the least distance while in the process of finishing. Second, use work slides or some other form of carrier so that when a workman complete his operation he drops the part always in the same place which must always be the most convenient place to his hand and if possible have gravity carry the part to the next workman. Third, use sliding assembly lines by which parts to be assembled are delivered at convenient intervals, spaced to make it easier to wok on them.

2. 完工的 机器设备 压铸 大量供应 火枪 兵工厂 任意地 结果

3. a) In the manufacturing industry, use of automatic assembly line can increase productivity and enable the product to maintain at a high level of quality.

   b) Automatic assembly process must be guaranteed with the interchangeable parts.

阅读材料 A 为网络选择通信协议

下面是 S7-200 CPU 所支持协议的总览。

-点对点接口（PPI）

-多对点接口（MPI）

-PROFIBUS

在开放系统互联（OSI）七层模式通信结构的基础上，这些通信协议在一个令牌环网络上实现。令牌环网络符合欧洲标准 EN 50170 中定义的 PROFIBUS 标准。这些协议是非同步的字符协议，有1位起始位、8位数据位、偶校验位和1位停止位。通信结构依赖于特定的起始字符和停止字符、源和目地站地址，帧长和数据校验和。在波特率一致的情况下，这些协议可以同时在一个网络上运行，并且互不干扰。

如果带有扩展模块 CP24-1 和 CP243-1 IT，那么 S7-200 也能运行在以太网上。

PPI 协议

PPI 是一种主–从协议：主站器件发送要求到从站器件，从站器件响应，如图 VA-1 所示。从站器件不发信息，只是等待主站的要求并对要求作出响应。主站靠一个 PPI 协议管理的共享链接与从站通信。PPI 并不限制与任意一个从站通信的主站数量，但是在一个网络中，主站的个数不能超过 32。

如果在用户程序中使能 PPI 主站模式，S7-200 CPU 在运行模式下可以作为主站。在使能 PPI 主站模式之后，可以使用网络读写指令来读写另外一个 S7-200。当 S7-200 作为 PPI 主站时，它仍然可以作为从站响应其他主站的请求。

PPI 高级允许网络设备建立一个设备与设备之间的逻辑连接。对于 PPI 高级，每个设备的连接个数是有限制的。S7-200 支持的连接个数见表 V A-1。

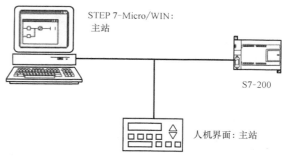

图 VA-1 点对点接口网络

表 VA-1 S7-200 CPU 和 EM277 模块的连接个数

模块	波特率	连接数
S7-200 CPU 通信口 0	9.6k、19.2k 或 187.5k	4
通信口 1	9.6k、19.2k 或 187.5k	4
EM277	9.6k 到 12M	6（每个模块）

所有的 S7-200 CPU 都支持 PPI 和 PPI 高级协议，而 EM277 模块仅仅支持 PPI 高级协议。

MPI 协议

MPI 允许主-主通信和主-从通信，如图 VA-2 所示。与一个 S7-200 CPU 通信，STEP 7-Micro/WIN 建立主-从连接。MPI 协议不能与作为主站的 S7-200 CPU 通信。

网络设备通过任意两个设备之间的连接通信（由 MPI 协议管理）。设备之间通信连接的个数受 S7-200CPU 或者 EM277 模块所支持的连接个数的限制。S7-200 支持的连接个数见表 VA-1。

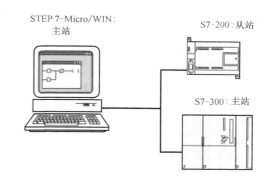

图 VA-2 多点接口网络

对于 MPI 协议，S7-300 和 S7-400 PLC 可以用 XGET 和 XPUT 指令来读写 S7-200 的数据。要得到更多关于这些指令的信息，参见 S7-300 或者 S7-400 的编程手册。

PROFIBUS 协议

PROFIBUS 协议通常用于实现与分布式 I/O（远程 I/O）的高速通信。可以使用不同厂家的 PROFIBUS 设备。这些设备包括简单的输入或输出模块、电机控制器和 PLC 等。

PROFIBUS 网络通常有一个主站和若干个 I/O 从站，如图 VA-3 所示。主站器件通过配置可以知道 I/O 从站的类型和站号。主站初始化网络使网络上的从站器件与配置相匹配。主站不断地读写从站的数据。

当一个 DP 主站成功配置了一个从站之后，它就拥有了这个从站器件。如果在网上有第二个主站器件，那么它对第一个主站的从站的访问将会受到限制。

TCP/IP 协议

通过以太网扩展模块（CP243-1）或互联网扩展模块（CP243-1 IT），S7-200 将能支持 TCP/IP 以太网通信。表 VA-2 列出了这些模块所支持的波特率和连接数。

若需更多信息，可参考 SIMATIC NET CP243-1 工业以太网通信处理器手册或 SIMATIC NET CP243-1 IT 工业以太网及信息技术通信处理器手册。

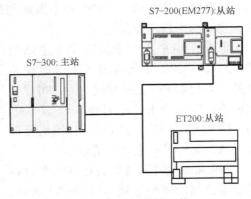

图 VA-3　现场总线网络

表 VA-2　以太网模块（CP243-1）和互联网模块（CP243-1 IT）的连接数

模块	波特率	连接数
以太网（CP243-1）模块	10～100M	8 个普通连接
互联网（CP243-1 IT）模块		1 个 STEP 7-Micro/WIN 连接

阅读材料 B　RobotStudio 简介

借助虚拟机器人技术进行离线编程，如同将真实的机器人搬到了您的 PC 机中！

离线编程是扩大机器人系统投资回报的最佳途径。借助 ABB 模拟与离线编程软件 RobotStudio，可在办公室内完成机器人编程，无需中断生产。机器人程序可提前准备就绪，提高整体生产效率。

借助 RobotStudio 提供的各种工具，可在不影响生产的前提下执行培训、编程和优化等任务，提升机器人系统的盈利能力，并让您获得多种利益：
- 降低风险
- 投产更迅速
- 换线更快捷
- 提高生产效率

RobotStudio 以 ABB VirtualController 为基础，与机器人在实际生产中运行的软件完全一致。因此，通过 RobotStudio 可执行十分逼真的模拟，采用车间中实际使用的真实机器人程序和配置文件。

阅读材料 C　计算机集成制造系统

一个计算机集成制造（CIM）系统通常被认为是包含所有生产活动的一个集成系统。生产活动囊括产品的计划、设计及控制。CIM 旨在将现有的计算机技术结合起来，以对整个生产过程进行管理和控制。许多公司以 CIM 为途径来建设未来的自动化工厂。

与传统的生产模式相比，CIM 的目的是以最低的成本，在最短的时间内将产品设计和材料转换成适销对路的商品。CIM 的应用把设计与制造之间原来的分离结合起来了。

CIM 与传统的加工车间的制造系统的区别在于计算机在制造过程中所起的作用。计算机集成制造系统基本上是由单一集成数据库联结的一个计算机系统网络。由于利用了数据库的信息、CIM 系统能够直接控制生产活动，记录结果并维护正确的数据。CIM 是集设计、制造、运送和财务功能为一体的计算机系统。

CIM 系统的一个主要组成部分是计算机辅助设计（CAD）系统。凡是利用计算机对工程设计进行开发、分析或修正的一切设计活动都属于 CAD 的范畴。由 CAD 系统完成的设计工作包括：

- 几何建模；
- 工程分析；
- 设计评审与评估；
- 自动绘图。

CIM 系统的另一主要部分是计算机辅助制造（CAM），采用 CAM 系统的一个重要原因是 CAM 系统为产品制造提供了一个数据库。然而，并非所有的 CAM 数据库都能与制造软件兼容。CAM 系统完成的任务有：

- 数控（NC）或计算机数控（CNC）编程；
- 计算机辅助工艺规划（CAPP）；
- 生产计划与安排；
- 刀具与夹具设计。

阅读材料 D 虚拟现实

虚拟现实系统是指能使一个或多个用户在计算机模拟环境中移动并能够与环境交互的系统。系统中各种各样的仪器能使参与者真实地感知和操纵虚拟的物体。这种自然的交互方式能使参与者在虚拟世界中产生沉浸感。这里的虚拟世界是由数学模型和计算机程序创造出来的。

虚拟现实仿真与其他计算机仿真的不同之处在于它需要特殊的接口设备，这种设备把虚拟世界的视觉、声音及感觉传递给用户。同时，这些设备记录参与者的语言和动作并发送给计算机仿真程序。

在未来，你的办公室可能已不再像小卧室而更像《星际旅行》中的全息驾驶舱。计算机科学家已经在做远程沉浸技术的实验。这项技术使我们在几百英里外能看到同事的办公室，就像我们与他们正处于同一个物理空间一样。

正在共同研究名为"国家远程沉浸初步"项目的计算机科学家已经展示出了一个原型系统，这个系统能使一个在 Chapel Hill 工作的科学家通过安装在他办公桌右边的两个互成直角的屏幕看到离他很远的同事。就好像从办公室的另一边透过玻璃朝里看一样。与视频会议不同，远程沉浸能提供与实物一样大小的三维图像。

在这个原型系统中，每个研究人员身边都有一排从不同角度监视他运动的数字照相机。研究人员还戴着头盔式跟踪设备和偏振眼镜，这种眼镜与看三维电影的眼镜一样。当研究人员头部移动时，他视野里的同事的位置也相应地变化。如果他向前倾，他的同事就会和他靠得更近，尽管实际上他们相隔几百英里。

科学家希望远程沉浸技术能开发出大量的其他应用，例如，地处乡村的病人能由遥远的城市的医生来治疗。

第六部分 文献检索

第 25 单元 信息检索简介

20 世纪 90 年代中期，随着电子文献的激增和互联网的迅速发展，文献检索的重点开始向电子数据库倾斜。为了适应网络时代的发展，在很多领域"信息检索"一词开始逐步取代"文献检索"。

信息检索（Information Retrieval），是指将信息按一定的方式组织和存储起来，并根据用户需要的信息找出有关的信息，所以它的全称又叫"信息的存储与检索"即 Information Storage and Retrieval，这是广义的信息检索。狭义的信息检索则仅指该过程的后半部分，即从信息集合中找出所需要的信息的过程，相当于人们通常所说的信息查寻（Information Search）。

文献信息检索是科学研究的向导。要进行有价值的科学研究，必须依赖文献检索，全面获取相关文献信息，及时了解各学科领域出现的新问题、新观点，以确定自己的研究起点和研究目标。科学研究首先需要通过课题调研来掌握资料，而文献检索则有助于掌握本课题研究的进展动态，开拓思路，避免重复劳动，提升研究水平。科研成果的评估也需要通过文献检索来进行鉴定，从而作出正确的结论。文献检索能力的高低，往往影响着科研成果的价值，无论是在课题确定阶段、科学实验阶段，还是成果总结阶段，文献检索都意义重大。可以说，文献检索对科研工作功不可没。

在课题确定阶段，必须查阅大量的文献资料，了解课题的历史与现状、前景与动向，把握前人做了什么，别人正在做什么，本课题还存在什么问题，有什么经验和教训。在充分调查的基础上，借鉴成功的经验、失败的教训和研究的方法，以便作出有别于他人的创新结论，制订出具体的科研计划。在科学实验阶段，在对客观规律的探索过程中，为了解决遇到的问题和困难，必须要借鉴前人的经验，以获得解疑排难的启示。在成果总结阶段，为了阐明研究成果的继承性和创造性，也必须广泛搜集有关论述，把他人的和自己的研究成果进行科学比较，作出客观评价，以充分证明其准确性和推广性。由此可见，在科学研究的各个环节，自始至终都需要借鉴、交流、积累和继承，都离不开对文献信息的利用。

所以，在科学研究过程中，做好文献检索不仅能够促进信息资源迅速、充分地开发利用，而且能够帮助科研人员继承和借鉴前人的成果，避免重复研究，少走弯路，加速研究进程。具体来讲，文献检索在为研究者服务方面的意义和作用主要有：（1）更具体地限制和确定研究课题及假设；（2）告诉研究者在本领域内前人或他人已经做了哪些工作；（3）提供一些可能对当前研究有价值的研究思路及方法；（4）对研究方案提出适当的修改意见，以免出现意想不到的困难；（5）把握在研究中可能出现的差错；（6）为解释研究结果提供背景材料。

文献可以分为零次文献、一次文献、二次文献、三次文献四种。零次文献是指未经过任何加工的原始文献，如实验记录、手稿、原始录音、原始录像、谈话记录等。一次文献（primary document）是指作者以本人的研究成果为基本素材而创作或撰写的文献。大部分期刊上发表的文章和在科技会议上发表的论文均属一次文献。二次文献（secondary document）是指文献工作者对一次文献进行加工、提炼和压缩之后所得到的产物。三次文献（tertiary document）主要包括大百科全书、词典等。

现代文献检索大多以计算机技术为手段,通过光盘和联机等现代检索方式进行信息检索。计算机信息检索是现代科技人员的一项基本功,这一能力的训练和培养对科技人员适应未来社会和跨世纪科研都极其重要。一个善于从电子信息系统中获取信息的科研人员,必定比不具备这一能力的人有更多的成功机会。在美国,交互网络检索专家已成为十大热门职业之一。

文献检索是一项实践性很强的活动,它要求我们善于思考,并通过经常性的实践,逐步掌握文献检索的规律,从而迅速、准确地获得所需文献。一般来说,文献检索可分为以下步骤:1)明确查找目的与要求。2)选择检索工具。3)确定检索途径和方法。4)根据文献线索,查阅原始文献。

目前,很多研究型高校都开设了"信息检索与利用"课。它是一门讲述基于网络环境下利用电子信息资源的课程。主要介绍因特网上各类电子信息资源的内容和检索技术,让学生认识各种类型的电子信息资源,系统了解和较为熟练地掌握各类数据库、电子期刊、电子图书、电子报纸以及因特网上的其他电子信息资源的检索方法。学生通过对这门课程的学习和实践,熟练地掌握查找和利用各种电子信息资源的技能,增强现代信息意识和研究能力、提高大学生获取文献以及自如地利用各种图书馆资源和网络信息资源的能力。

练习答案

1. a)狭义的文献检索仅是指人们通常说的搜索信息。是

b)在科学研究的各个环节,都离不开对文献信息的利用。是。

c)文献可以分为零次文献、一次文献、二次文献三种。不对,应该是四种.

2. a)yes

3. a)Modern literature retrieval is mainly based on computer technology, through the CD, online network, and other modern retrieval way.

b)At present, many research universities open the course named Information Retrieval and Use.

第 26 单元　专业文献的检索

文献检索可分手工检索和计算机检索两种。计算机信息检索的优点在于速度快、耗时少、查阅范围广,甚至可以查到国外刚刚出版的期刊论文的信息。同时,其检索内容的专指性很强。计算机信息检索的缺点在于追溯时间受到一定限制,检索费用比较昂贵,检索的时间范围也有一定的限制。计算机信息检索主要适用于已经数字化的近期文献信息和动态性信息的查找。手工检索的优缺点差不多正好与计算机信息检索的相反。手工检索的优点是检索时间和检索范围都不受限制。但是手工检索耗时多、效率低、检索入口少,因此查找效果往往不如计算机信息检索的好。手工检索主要适用于纸质印刷书刊文献,特别是早期文献信息的查找。

专业文献的检索需要通过比较专业的方法来实现,其检索途径多种多样,一般包括:

1. 按著者检索

许多检索系统备有著者索引,专利文献检索系统则有专利权人索引。著者可以是作者、编者、译者或专利权人;著者可以是个人,也可以是单位或机构。

2. 按题名检索

顾名思义，题名是指书名、刊名、篇名等。绝大部分检索系统中都提供按题名字顺检索的途径，如书名目录和刊名目录等。

3. 按分类检索

分类是指将各种概念按照学科、专业性质进行区分和系统排列，让用户能够通过这个系统查找自己所需要的文献；分类比较能体现学科系统性，反映学科与事物的隶属、派生与平行的关系，便于我们从学科所属范围来查找文献资料，并且可以起到"触类旁通"的作用。

4. 按主题检索

主题检索的优点在于能够集中反映一个主题的各方面文献资料，因而便于读者对某一问题、某一事物和对象作全面系统的专题性研究。我们通过主题目录或索引，即可查到同一主题的各方面文献资料。

5. 按引文检索

文献所附参考文献或引用文献，是文献的外表特征之一。利用这种引文而编制的索引系统，称为引文索引系统。研究者可以从被引论文去检索相关文献资料。

6. 按序号检索

有些文献有特定的序号，如专利号、报告号、合同号、标准号、国际标准书号和刊号等。文献序号对于识别一定的文献，具有明确、简短、唯一性的特点。依此编成的各种序号索引可以提供按序号自身顺序检索文献信息的途径。

7. 按代码检索

代码检索是指利用事物的某种代码编成的索引，如分子式索引、环系索引等，按特定代码顺序进行检索。

8. 按专项检索

有些文献中包含着名词术语、地名、人名、机构名、商品名、生物属名、年代等特定信息，利用这些信息进行检索，可以解决某些特别的问题。

中国期刊网是研究者使用得最广泛而频繁的网络平台。全面了解中国期刊网，对于提高检索效率和质量有重要意义。

中国期刊网专题全文数据库共包括理工 A、理工 B、理工 C、农业、医药卫生、文史哲、经济政治与法律、教育与社会科学、电子技术与信息科学九个专辑。检索手段包括13个检索字段，它们分别是篇名字段、作者字段、关键词字段、主题词字段、机构字段、中文刊名字段、中文摘要字段、引文字段、基金字段、全文字段、年份字段、期数字段和 ISSN 字段。其中，篇名、文摘、关键词、主题词和全文字段属于基本字段。中国期刊网专题全文数据库共有三种检索方法。它们是：1）分类导航。利用系统提供的专辑栏目及其展开的分类导航树查找相关文献，适用于对一个学科的文献做全面调查的场合。2）初级检索。利用检索项对某个指定字段进行检索。3）高级检索。是指设置多个检索条件和逻辑组合关系，查找同时满足这些条件和逻辑关系的文献。中国期刊网专题全文数据库设有二次检索功能，可以在前次检索结果的范围内再次进行查找，以达到缩小检索范围，使检索结果逐步接近课题要求的目的。不管是初级还是高级检索界面，只要检索结果的篇数允许，二次检索可以反复执行，直到满足研究要求为止。中国期刊网设有专题全文数据库的全文浏览器功能，其作用是为了便于浏览这个数据库的全文。全文浏览器可以在该数据库首页上下载。

练习答案

1. a) 著者可以是个人，也可以是单位或机构。是。

b) 中国期刊网是研究者通常使用得最广泛而频繁的网络平台。是。

c) 不是。中国期刊网专题全文数据库共包括九个专辑。

d) 是。中国期刊网的检索方法包括 13 个检索字段。

2. 在中国期刊网中，篇名、文摘、关键词、主题词和全文字段属于基本字段。

3. a) We can get all literature of the same theme through the subject catalogue or index.

b) Whether it is the primary or advanced search interface, as long as retrieval result number permitted, the secondary retrieval may operate repeatedly, until satisfying the research request.

4. 略。

5. 请用你常用的两种搜索引擎查找专业资料，写清过程和结果，并对检索结果作出评价。过程略。

Appendix B 常用专业词汇

A

a bank of	一系列；一排，一束	（ⅤD）
a host of	一大群	（ⅤD）
abbreviation [əˌbriːviˈeiʃən]	n. 缩写字，缩写式；缩写，省略；缩短	（21）
abrasive [əˈbreisiv]	n. 研磨剂，磨料（具）；adj. 研磨的	（1）
abrasive wheel [əˈbreisiv]	砂轮	（ⅡB）
account for	占；计算出；解释，说明	（24）
achieve [əˈtʃiːv]	vt. 完成，实现；达到，赢得	（21）
actuator [ˈæktjueitə]	n. 激励者；[电脑] 执行机构；[电]（电磁铁）螺线管；[机] 促动器	（ⅣB）
adhesive [ædˈhiːsiv]	n. 黏结剂，黏胶剂	（1）
advantage [ədˈvɑːntidʒ]	n. 有利条件，优点，优势；利益，好处	（26）
algebra [ˈældʒibrə]	n. 代数学，代数	（10）
algebraic [ˌældʒiˈbreik]	adj. 代数的	（2）
algorithm [ˈælgəriðəm]	n. 运算法则；演算法；计算程序	（ⅣA）
alignment [əˈlainmənt]	n. 成直线，（直线）对准	（5）
allowance [əˈlauəns]	n. 容差	（ⅠB）
amplification [ˌæmpləfiˈkeiʃən]	n. 扩大	（10）
analog [ˈænəlɔg]	n. 类似物；同源语；模拟；adj.（钟表）有长短针的；模拟的	（10）
anneal [əˈniːl]	n. 退火	（3）
apron [ˈeiprən]	n. 溜板箱，挡板	（5）
arc welding	（电）弧焊	（22）
archive [ˈɑːkaiv]	n. 存档；档案文件，档案室	（19）
arsenal [ˈɑːsənəl]	n. 兵工厂，军械库	（24）
as long as = so long as	只要	（26）
as the name suggests	顾名思义	（26）
assignment [əˈsainmənt]	n. 分配；任务，工作，作业；指定，委派	（Ⅳ）
asynchronous [eiˈsiŋkrənəs]	adj. 异步的	（ⅤA）
attention [əˈtenʃən]	n. 注意，专心，留心	（12）
automatically [ˌɔːtəˈmætikəli]	adv. 自动地；无意识地，不自觉地，机械地	（21）
Auto-Tuning Control Panel	自动整定控制面板	（14）

B

backlash [ˈbæklæʃ]	n. 轮齿隙；反斜线；后座；后冲	（ⅡA）

be at right angles to	与……成直角	(7)
be formed from…	由……组成	(1)
be used for…	用于……	(1)
behavior [biˈheivjə]	n. 行为，举止；态度	(11)
belt [belt]	n. 层，界	(6)
bilateral [baiˈlætərəl]	adj. 双向的，双边的	(ⅠB)
blend [blend]	vt. & vi. （使）混合，（使）混杂；n. 混合物	(12)
Boolean logic	布尔逻辑	(14)
boundary [ˈbaundəri]	n. 分界线；范围；（球场）边线	(ⅣB)
brine [brain]	n. 盐水	(3)
brittle [ˈbritl]	adj. 易碎的	(3)
built-in [ˌbiltˈin]	adj. 嵌入的；内置的；固有的	(14)
bulk [bʌlk]	n. 整体，容量	(6)
burglar [ˈbə:glə]	n. 窃贼，破门盗窃者	(11)
bushing [ˈbuʃiŋ]	n. 衬套；轴衬；轴瓦；[电]（绝缘）套管	(ⅡC)

C

cabinet [ˈkæbinit]	n. 内阁；柜橱；小房间；展览艺术品的小陈列室	(ⅣB)
cam [kæm]	n. 凸轮，偏心轮；样板，靠模，仿形板	(ⅡC)
cam-lock	偏心夹，凸轮锁紧	(5)
capital-intensive [ˌkæpitlinˈtensiv]	n. 资金集约型	(20)
carriage assembly	溜板箱组件	(5)
carriage [ˈkæridʒ]	n. 拖板，大拖板	(5)
carry out	完成，实现，贯彻，执行	(23)
cast iron	铸铁	(1)
category [ˈkætəgəri]	n. 范畴	(17)
cemented carbide	硬质合金钢	(ⅣD)
center [ˈsentə]	n. 顶尖	(5)
ceramic [siˈræmik]	n. 陶瓷，陶瓷制品	(1)
chamber [ˈtʃeimbə]	n. 油腔	(15)
channel [ˈtʃænl]	n. 槽钢	(4)
chatter [ˈtʃætə]	vi. 振动	(8)
checksumn [ˈtʃeksʌm]	n. 校验和	(ⅤC)
chisel [ˈtʃizəl]	n. 钻头横刃	(8)
chuck [tʃʌk]	n. 卡盘	(5)
close tolerance	紧公差	(ⅠB)
closeness [kləuznis]	n. 接近（程度）	(2)
clutch [klʌtʃ]	n. 离合器，联轴器；夹紧装置	(ⅡC)

CMOS (Complementary Metal Oxide Semiconductor)	互补金属氧化物半导体	(13)
CNC (Computer Numerical Control)	计算机数控	(20)
coherent [kəu'hiərənt]	adj. 连贯的，一致的	(ⅤC)
coil [kɔil]	n. （一）卷，（一）圈；盘卷之物；线圈；簧圈；盘管	(9)
cold forming	冷成形，冷态成形，冷作成形	(4)
cold-working	冷加工	(ⅠC)
collision [kə'liʒən]	n. 碰撞；冲突；（意见，看法）的抵触	(22)
column-and-knee	升降台式	(ⅡD)
commercialization	n. 商品化，商业化	(21)
compact [kəm'pækt]	adj. 紧凑的，紧密的，简洁的	(15)
compatible [kəm'pætəbl]	adj. 适合的；相容的	(13)
compensate ['kɔmpenseit]	v. 偿还，补偿，付报酬	(ⅡA)
competitive [kəm'petitiv]	adj. 竞争的；竞争性的；好竞争的	(21)
complex math operations	复杂数学运算	(14)
component [kəm'pəunənt]	n. 组成部分，部件，元件	(16)
composite ['kɔmpəzit]	adj. 合成的，复合的，混合物	(1)
composition [kɔmpə'ziʃən]	n. 构图；构成，成分	(11)
comprehensive [ˌkɔmpri'hensiv]	adj. 综合的	(17)
configuration [kən'figju'reiʃən]	n. 外形，轮廓	(6)
conical ['kɔnikəl]	adj. 圆锥的，圆锥形的	(4)
consciousness ['kɔnʃəsnis]	n. 意识，意念	(25)
consumption [kən'sʌmpʃən]	n. 消费，消耗；消费［耗］量	(10)
contour ['kɔn'tuə]	n. 轮廓（线），外形	(7)
contribution [kɔntri'bju:ʃən]	n. 捐助物，贡献	(9)
conversion [kən'və:ʃən]	n. 变换，转化	(10)
corrosion [kə'rəuʒən]	n. 腐蚀	(1)
corrosion resistance	耐（腐）蚀性，耐蚀力，抗腐（蚀）性	(22)
co-worker	n. 合作者	(ⅤD)
crank [kræŋk]	n. 曲轴	(ⅡC)
critical ['kritikəl]	adj. 危急的；决定性的；［物］临界的	(19)
cross slide	中拖板	(5)
crude [kru:d]	adj. 天然的，未经加工的	(1)
cubicle ['kju:bikəl]	n. 小卧室	(ⅤD)
curriculum [kə'rikjuləm]	n. 课程	(17)
cutting ['kʌtiŋ]	v. 切削	(20)

D

database ['deitə₁beis]	n. 资料库，数据库	(25)
datum ['deitəm]	n. 基准（点，线，面）	(ID)
Daylight Savings time	n. 夏令时	(14)
deburring [di'bə:riŋ]	n. 修边，除去毛刺	(22)
deep drawing	深拉，探冲（压）	(4)
density ['densiti]	n. 密集，稠密；〈物〉〈化〉密度	(11)
depict [di'pikt]	vt. 画，叙述	(2)
designate ['dezigneit]	vt. 标出，把…定义为…	(2)
diagnostic [₁daiəg'nɔstik]	adj. 诊断的，判断的；特征的	(14)
diagram ['daiəgræm]	n. 图解，简图，图表	(9)
directive [di'rektiv]	n. 指令；<美>命令，训令，指令；方针	(ⅣB)
discipline ['disiplin]	vt. 训练，训导；n. 训练，学科	(12)
dismantle [dis'mæntl]	vt 分解（机器），拆开，拆卸	(ⅡC)
distinguished [dis'tiŋgwiʃt]	adj. 卓越的，杰出的	(24)
distribution [distri'bju:ʃən]	n. 分发，分配	(9)
dress press	钻床	(8)
drill press	钻床，立式钻床	(20)
ductile ['dʌktail]	adj. 柔软的，易延展的	(3)
ductility [dʌk'tiliti]	n. 韧性，可延展性	(1)
dynamically [dai'næmikəli]	adv. ［计］动态地	(19)

E

eccentric [ik'sentrik]	adj. 偏心的，不同圆心的	(7)
ejection [i'dʒekʃən]	n. 挤出，抛出	(8)
Electrical Engineering	电气工程	(17)
eliminate [i'limineit]	vt. 排除；消除；除掉；	(ⅣA)
embedded [em'bedid]	adj. 植入的，深入的，内含的	(13)
emission [i'miʃən]	n. 排放，辐射；排放物，散发物（尤指气体）；（书刊）发行	(22)
empirical [em'pirikəl]	adj. 经验（主义）的	(2)
encompass [en'kʌmpəs]	vt. 包围，环绕，包含或包括某事物	(ⅤC)
end effector	末端执行器	(23)
end-feed	纵向定程进给	(6)
engage [in'geidʒ]	v. 啮合	(5)
enhancement [in'hɑ:nsmənt]	n. 增强；增加；提高；改善	(14)
environment [in'vaiərənmənt]	n. 环境，四周状况	(25)
equivalent-to	相当于	(25)

even parity		奇偶校验	(ⅤA)
excess	['ekses]	adj. 超重的，过量的；额外的；n. 超过，过多之量	(9)
extremely	[iks'tri:mli]	adv. 极端地；极其地；非常地	(10)

F

face plate		面板，花盘	(5)
facilitate	[fə'siliteit]	vt. 使容易；促进，帮助	(26)
fiberglass	['faibəglɑ:s]	n. 玻璃纤维	(1)
filter	['filtə]	n. 过滤，过滤器；vt. & vi. 透过，过滤	(10)
finished	['finiʃt]	adj. 完美的，精加工的，完工的	(2)
finishing	['finiʃiŋ]	n. 带式磨光，饰面，表面修饰，擦光	(20)
fixture	['fikstʃə]	n. 夹具，夹紧装置	(2)
flexibility	[fleksi'biliti]	n. 柔韧性；机动性，灵活性	(12)
flexible	['fleksəbl]	adj. 灵活的；易弯曲的；柔韧的；易被说服的	(13)
Flexible Manufacturing Systems (FMS)		柔性制造系统	(20)
fluctuate	['flʌktjueit]	vi. 波动，涨落，起伏	(11)
flute	[flu:t]	n. 凹槽，出屑槽	(8)
force fit		压入配合．压紧配合	(ⅠB)
forego	[fɔ:'gəu]	v. 先行，在前，前面	(2)
forging machine		锻造机	(20)
formability	['fɔ:mə'biliti]	n. 可成型性	(1)
frame	[freim]	n. 框架；眼镜框；组织；边框	(ⅤA)
frequent	['fri:kwənt]	adj. 时常发生的，频繁的；惯常的；习以为常的	(26)
full annealing		完全退火	(3)
functionality	[ˌfʌŋkʃə'næliti]	n. 功能；功能性；设计目的	(19)

G

gap	[gæp]	n. 缺口，裂口，间隙，缝隙，差距	(4)
generation	[dʒenə'reiʃən]	n. 同时代的人，一代人，一代	(12)
geometric	[ˌdʒi:ə'metrik]	adj. 几何学（的）	(ⅤA)
grade (quality) of tolerance		公差等级	(2)
gradually	['grædjuəli]	adv. 逐步地，渐渐地	(25)
grinder	['graində]	n. 磨床，研磨机，磨工	(20)
grinding	['graindiŋ]	n. 磨削	(20)
grinding	['graindiŋ]	n. 磨削	(2)
GUI (Graphical User Interface)		图形用户界面	(19)
guidance	['gaidəns]	n. 导航，指引；领导；制导，向导	(23)
guideline	['gaidlain]	n. 指导方针	(ⅣB)

H

hand plane	手刨，木工刨	(ⅡB)
headstock assembly	主轴箱组件	(5)
headstock ['hedstɔk]	n. 主轴箱，床头箱	(5)
heat treatment	热处理	(3)
heat-resistant material	耐热材料	(1)
heat-resistant steel	耐热钢	(ⅠA)
helical gear	斜齿轮	(ⅡA)
helical ['helikəl]	adj. 螺旋形的；螺旋线的	(ⅡA)
helices ['helisi:z]	n. （pl）螺杆，螺旋状（之物）	(ⅡD)
herringbone ['heriŋbəun]	n. 交叉缝式，人字形；adj. 人字行的；v. （使）成箭尾形	(ⅡA)
high-speed steel	高速钢	(ⅠA)
HMI (Human Machine Interface)	人机界面	(19)
hole-basis system	基孔制	(2)
holo-deck	n. 全息驾驶舱	(ⅤD)
hone [həun]	n. & v. 磨	(6)
honing ['həuniŋ]	n. 珩（搪）磨	(2)
hopper ['hɔpə]	n. 漏斗，料斗	(24)
hot forming	热成型	(4)
hot-working	热加工	(ⅠC)
human-computer interfacing	人机界面	(23)
hydraulic [hai'drɔ:lik]	adj. 液压的	(15)
hydraulic cylinder	油缸	(ⅣC)
hydraulic driver	静压传动	(ⅣC)
hydraulic motor	液压马达	(ⅣC)

I

identify [ai'dentifai]	vt. 识别，认出；确定；使参与；把……看成一样	(ⅣB)
imminent ['iminʃnt]	adj. 危急的，急迫的	(24)
impetus ['impitəs]	n. 推动，促进，刺激	(11)
implement ['implimənt]	vt. 使生效，贯彻，执行；n. 工具，器具，用具	(10)
in advance	在前头；预先，事先	(22)
inability ['inə'biliti]	n. 无能，无力	(4)
incline [in'klain]	v. 使倾斜；赞同；喜爱	(ⅡA)
incorporate [in'kɔ:pəreit]	adj. 合并的，结社的；v. 合并，组成公司	(20)
independence [ˌindi'pendəns]	n. 独立，自主；	(ⅣB)
index ['indeks]	vt. 分度	(6)

英文	中文	出处
in-feed	横向进给	(6)
ingot ['iŋgət]	n. 工业纯铁	(ⅠC)
inhibit [in'hibit]	v. 抑制；禁止	(ⅣA)
initiative [i'niʃiətiv]	n. 主动（性），第一步	(ⅤD)
input filters	输入滤波器	(14)
inseparable [in'sepərəbl]	adj. 分不开的，不可分离的	(25)
inspection [in'spekʃən]	n. 检查，探伤，验收；目测；验证，校验	(23)
instruction set	指令集	(13)
In-System Programmable Flash Memory	在系统可编程的 Flash 只读存储器	(13)
integral ['intigrəl]	adj. 整（数，体）的，完整的	(7)
integrate ['intigreit]	vt. 集成，使一体化，积分 v. 结合	(20)
integrated ['intəˌgratid]	adj. 综合的，完整的	(23)
integration [ˌinti'greiʃən]	n. 整合，完成；集成	(21)
integrity [in'tegriti]	n. 正直，诚实；完整；[计算机] 保存；健全	(ⅤA)
interchangeable [intə'tʃeindʒəbl]	adj. 可互换的，通用的	(24)
interface ['intəfeis]	n. 界面；<计>接口；交界面	(19)
interference ['intə'fiərəns]	n. 过盈，	(2)
interference fit	干涉配合，静配合	(ⅠB)
interlocking ['intə'lɔkiŋ]	adj. 可联动的，互锁的	(7)
Interrupt Event	n. 中断事件	(14)
interval ['intəvəl]	n. [军事] 间隔；间隔时间；（戏剧、电影或音乐会的）幕间休息	(14)
invoke [in'vəuk]	vt. 祈求；提出或授引……以支持或证明；唤起；引起	(ⅣA)
irregularity [i'regju'læritiː]	n. 不规则（均匀，对称）性	(ⅠD)

J

| job-shop | 加工车间，机修车间 | (7) |

K

| kinetic energy | 动能 | (ⅣC) |

L

land [lænd]	n. 刀棱面，齿刃	(7)
lap [læp]	n. & v. 研磨，抛光	(6)
lapping ['læpiŋ]	n. 研磨，抛（磨）光	(2)
lathe [leið]	n. 车床	(5)

leadscrew ［li:dskru:］	n.	丝杠，导（螺）杆	(5)
leave off ［'li:vɔf］	vt.	停止（做）某事，戒掉	(ⅣA)
lightweight ［'laitweit］	adj.	轻量的，薄型的；n. 轻量	(10)
load-bearing		承载	(1)
log ［lɔg］	vt.	把……记入航海日志；航行（…距离）	(19)
logic ［'lɔdʒik］	n.	逻辑（学），逻辑性	(10)
lubricant ［'lu:brikənt］	n.	润滑剂	(1)
lubrication ［ˌlu:bri'keiʃən］	n.	润滑	(ⅡA)

M

machinability ［məˌʃi:nəbiliti］	n.	切削性，机械加工性	(3)
machine-tool		机床	(20)
machining centers		加工中心	(18)
management ［'mænidʒmənt］	n.	管理，经营，处理；管理部门；经理部	(21)
manifold ［'mænəˌfəuld］	adj.	多种多样的；多方面的；有多种形式的；有多种用途的	(19)
master-slave protocol		主从协议	(ⅤA)
material-handling		物料输送，原材料处理	(20)
mating ［'meitiŋ］	n. & adj.	配合（的），相连（的）	(2)
mean roughness index		轮廓算术平均偏差	(ⅠD)
mechanism ［'mekənizəm］	n.	机械装置，机构，机制	(ⅣD)
medium ［'mi:diəm］	adj.	中等的，中间的	(3)
microcontroller ［ˌmaikrəkən'trəulə］	n.	微控制器	(13)
microelectromechanical		微机电	(17)
micrometer ［maikrɔ'mitə］	n.	微米，千分尺	(2)
military ［''militəri］	adj.	军事的，军用的	(18)
milling ［'miliŋ］	n.	铣削	(18)
milling machine		铣床	(20)
module ［'mɔdju:l］	n.	组件，单元；（航天器的）舱	(21)
molecular ［məu'lekju:lə］	adj.	分子的	(1)
molecule ［'mɔlikju:l］	n.	分子	(1)
molybdenum ［mə'libdinəm］	n.	钼	(ⅠA)
monolithic ［ˌmɔnə'liθik］	adj.	独块巨石的；整体的；庞大的	(13)
mount ［maunt］	v.	增加；上升 vt. 安装，架置；镶嵌，嵌入；准备上演	(22)
multi-disciplinary		多学科	(23)
multiplexed ［'mʌltiˌpleksid］	adj.	多路复用的	(13)
multipurpose ［ˌmʌlti'pə:pəs］		多种用途的，多目标的	(ⅣD)

musket [ˈmʌskit]	n. 火枪	(24)

N

nickel [ˈnikəl]	n. 镍	(1)
nodular [ˈnɔdjulə]	adj. 球(团,粒)状的	(5)
nonvolatile [nɔnˈvɔlətail]	adj. 非易失性的	(13)
normalizing [ˈnɔːməˌlaiziŋ]	n. 正火	(3)
nut [nʌt]	n. 螺母	(ⅡC)

O

objective law	客观规律	(25)
offline [ˈɔflain]	adj. adv. 未连线的(地);脱机的(地);离线的(地)	(22)
oil pressure pump	油泵	(ⅣC)
onlooker [ˈɔnˌlukə]	旁观者	(ⅣD)
Open System Interconnection	开放系统互连	(ⅤA)
operator [ˈɔpəreitə]	n. (机器、设备等的)操作员,机务员	(19)
optimal [ˈɔptəməl]	adj. 最佳的,最优的	(12)
optimization [ˌɔptimaiˈzeʃən]	n. 最佳化,最优化	(12)
orifice [ˈɔːrəfis]	n. 孔,节流孔	(4)
original [əˈridʒənl]	adj. 最初的,本来的;原始的	(25)
originate [əˈridʒineit]	v. 起源于,来自,产生;vt. 创造,创始,开创;发明	(12)
oscillator [ˈɔsileitə]	n. 振荡器	(13)
outboard [ˈautbɔːd]	adj. 外置的	(ⅢD)
overlap [ˌouvəˈlæp]	v. (与某物)交叠,重叠,重合	(ⅡA)

P

pallet [ˈpælit]	n. 托盘,货盘	(20)
parameter [pəˈræmitə]	n. 参数,系数	(6)
parameterization [ˌpærəˌmitəriˈzeiʃən]	n. 参数化,参数化法	(14)
participation [pɑːtisəˈpeiʃən]	n. 参加,参与	(11)
partition [pɑːˈtiʃən]	vt. 分开,隔开;区分;分割	(ⅣB)
patent [ˈpeitənt]	n. 专利权;执照;专利品;vt. 授予专利;取得……的专利权	(26)
peak-to-valley height	波峰极点	(ⅠD)
pennsylvania [pensilˈveninjə]	宾夕法尼亚洲(美国州名)	(18)
performance [pəˈfɔːməns]	n. 演出,表演,性能,工作情况	(12)

peripheral [pəˈrifəiəl]	adj. 圆周的、周边的，外部的 n. 外部设备	(7)
permissible [pəˈmisəbl]	adj. 可允许的，许可的	(2)
phenomenon [fiˈnɔminən]	n. 现象	(9)
pictorially [pikˈtɔːriːəli]	adv. 用（插）图，如绘成图画	(2)
PID (propotional-integral-derivative)	比例积分微分	(14)
pinion [ˈpinjən]	n. 小齿轮	(ⅡA)
pinout [ˈpinaut]	n. 引出线	(13)
piston [ˈpistən]	n. 活塞	(15)
pitch [pitʃ]	n. （齿轮）节距	(ⅡA)
planer [ˈpleinə]	n. 龙门刨床	(ⅡB)
planning [ˈplæniŋ]	n. 刨（削，平）	(2)
plant [plɑːnt]	n. 设备；工厂	(19)
plywood [ˈplaiwud]	n. 夹板	(1)
polymer [ˈpɔlimə]	n. 聚合物	(1)
polymerization [ˈpɔliməraiˈziʃən]	n. 聚合	(1)
post-machining [pəustməˈʃiːniŋ]	n. 后续（期）加工	(2)
potential [pəˈtenʃəl]	adj. 潜在的，有可能的；n. 潜力，潜势，可能性	(9)
powder metallurgy	粉末冶金	(20)
PPI (Point to point interface)	n. 点对点接	(14)
predetermine [priːdiˈtɜːmin]	vt. & v. 预先裁定	(11)
predict [priˈdikt]	vt. & v. 预言；预测；预示	(9)
predominantly [priˈdɔminəntli]	adv. 占主导地位地；显著地；占优势地	(22)
preference [ˈprefərəns]	n. 优先选择	(15)
process annealing	低温退火；中间退火	(3)
professional [prəˈfeʃənəl]	adj. 职业性的，非业余性的	(26)
PROFIBUS abbr. (process field bus)	过程现场总线	(ⅤA)
profitability [ˌprɔfitəˈbiləti]	n. 获利（状况），盈利（情况）；收益性；利益率；有利	(22)
programmer [ˈprəuˌgræmə]	n. 程序设计者；程序设计器	(13)
protocol [ˈprəutəˌkɔːl]	n. 礼仪；（数据传递的）协议；科学实验报告（或计划）	(ⅤA)
protrude [prəˈtruːd]	v. （使）伸，突出	(5)
pulse catch	脉冲捕获	(14)
punching [pʌntʃ]	v. 凿孔，冲板，冲压	(20)

Q

| quenching [kwentʃiŋ] | n. 淬火 | (3) |

quill [kwil] n. 活动套筒 (5)

R

racture ['fræktʃə] n. 断裂 (4)
radial aim drill press 摇臂钻床 (8)
radius ['reidjəs] n. 半径（距离）；用半径度量的圆形面积；半径范围 (22)
ram-type 滑枕式 (ⅡD)
ratable ['reitəbl] adj. 可评价的，可估价的，按比例的 (ⅡA)
ream [riːm] vt. 铰孔，铰大……的口径 (2)
recipe ['resəpi] n. 烹饪法；食谱；方法；秘诀 (14)
recognition [ˌrekəg'niʃən] n. 识别，认得 (23)
reconfigurable adj. 可重构的；重新组态 (21)
rectify ['rektifai] vt. 消除，改正 (8)
reduce [ri'djuːs] vt. 减少；削减；缩小；使化为，使变为 (26)
reduction [ri'dʌkʃən] n. 减少；降低 (22)
reference ['refərəns] n. 参考，参照；参考书目；介绍信 vt. 引用 (25)
regulating valve 调节阀 (ⅣC)
reliable [ri'laiəbl] adj. 可靠的，可信赖的 (10)
relief valve 安全阀 (ⅣC)
remoteness [ri'məutnis] n. 偏（疏）远 (2)
removal [ri'muːvəl] n. 移走，脱掉 (9)
repeatability 可重复性，反复性，再现性 (ⅣD)
reposit [ri'pɔzit] vt. 贮藏，使复位 (9)
reposition [ˌriːpə'ziʃən] vt. 改变……的位置 (8)
requirement [ri'kwairmənt] n. 需要；必需品；要求；必要条件；规定 (26)
resistance [ri'zistəns] n. 抵抗，反抗，抵抗能力，电阻，热阻 (9)
resonance ['rezənəns] n. 回响，回荡；洪亮；共鸣 (12)
respectively [ris'pektivli] adv. 各自地，各个地，分别地 (9)
resume [ri'zjuːm] vt. 重新取得；再占有；取回 (ⅣA)
retraction [ri'trækʃən] n. 收回 (22)
retrieval system n. 回收系统 (20)
retrieve [ri'triːv] vt. 取回；恢复；[计] 检索；重新得到 (19)
rivet ['vivit] n. 铆钉 v. 铆接，铆 (ⅡC)
robust [rəu'bʌst] adj. 精力充沛的；坚定的；粗野的，粗鲁的；需要体力的 (22)

S

scaling ['skeiliŋ] v. 剥落，氧化起皮 (ⅠC)
scheme [skiːm] n. 计划，方案 (6)

score [skɔː]	v. 刻划,研刻	(8)
self-hardening steel	自硬钢	(ⅠA)
sensitivity [ˌsensiˈtiviti]	n. 敏感;敏感度	(16)
serpentine robotic arm	蜿蜒(蛇行)机器手臂	(23)
servomechanism [ˈsəːvənˈmekənizəm]	伺服系统	(18)
shaft-basis system	基轴制	(2)
shaper [ˈʃeipə]	n. 牛头刨床	(ⅡB)
shaping [ˈʃeipiŋ]	n. 成形加工	(2)
shatter [ˈʃætə]	v. 粉碎,破坏	(1)
shearing [ˈʃiəriŋ]	v. 剪切	(20)
shield [ʃiːld]	vt. 保护;掩护;庇护;给……加防护罩	(22)
shim [ʃim]	n. (薄)垫	(7)
shock resistancec	抗冲击性	(1)
shrinkage [ˈʃriŋkidʒ]	n. 收缩	(4)
shut down	关闭;停工	(22)
simulation [ˌsimjəˈleiʃən]	n. 模仿,模拟;[生]拟态,拟色;假装;装病	(22)
simultaneously [siməlˈteiniəsly]	同时发生地;同步地	(18)
single-point tool	单刃刀具	(ⅡB)
sinter [ˈsintə]	使烧结	(ⅣD)
slab [slæb]	n. 平面,平板	(7)
sleeplessly [ˈsliːplisli]	adj. 无阶级地	(15)
slot [slɔt]	n. 槽(沟),(裂)缝	(7)
soaking pit	均热坑	(ⅠC)
solenoids [ˈsəulinɔid]	n. 螺线管	(ⅣB)
sophistication [səˌfistiˈkeiʃən]	n. 老练,成熟,精致,世故	(20)
spectrum [ˈspektrəm]	n. 范围;系列,范围,幅度	(19)
spheroidizing [ˈsfiərɔidaiziŋ]	n. 球化退火	(3)
spindle [ˈspindl]	n. 主轴,轴	(18)
spline [splain]	n. 花键(轴)	(6)
split [split]	v. 劈开,(使)裂开;分裂,分离 n. 裂开,裂口,裂痕	(ⅡA)
spraying [ˈspreiiŋ]	n. 喷雾	(22)
spur [spəː]	n. [建]凸壁;支撑物	(ⅡA)
spur gear	正齿轮	(ⅡA)
squeeze [skwiːz]	v. 挤,压	(ⅠC)
stability [stəˈbiliti]	n. 稳定性	(17)
stabilize [ˈsteibəlaiz]	vt. & v. (使)稳定,(使)稳固;使稳定平衡	(ⅣA)
stage [steidʒ]	n. (进展的)阶段;时期	(25)

英文	音标	词性释义	出处
staggered	['stægəd]	adj. 交错的	(7)
stainless	['steinlis]	adj. 不锈的	(ⅠA)
stanford	['stænfəd]	n. 斯坦福	(17)
Star Trek		《星际旅行》	(ⅤD)
stepped	[stept]	adj. 分级的，极形的	(6)
stepped-vee pulley		台阶式V形槽带轮	(8)
stiffness	['stifnis]	n. 硬度，刚度	(1)
straddle	Pstraedl	adj. 跨式的	(7)
strain hardening		应变硬化，加工硬化 冷作硬化	(4)
stress relieving		去应力退火	(3)
stringent	['strindʒənt]	adj. 严格的；迫切的；（货币）紧缩的	(22)
strobe	[strəub]	n. 闸门，起滤波作用	(13)
subdivide	[ˌsʌbdi'vaid]	v. 细（区）分，再（划）分	(2)
successive	[sək'sesiv]	adj. 继承的，连续的	(20)
suffix	['sʌfiks]	v. 满足（…的需要）	(2)
superstructure	['sjuːpəˌstrʌktʃə]	n. 上部结构	(4)
surface roughness		表面粗糙度	(ⅠD)
symbol	['simbəl]	n. 象征，标志	(10)
symbolic	[sim'bɔlik]	adj. 象征的，象征性的	(ⅣB)
synthesis	['sinθisis]	n. 综合，综合法；〈化〉合成	(12)

T

英文	音标	词性释义	出处
tableware	['teib(ə)lweə]	n. 餐具	(1)
tailstock	['teilstɔk]	n. 尾座，尾架	(5)
tele-immersion		远程投入，远程参与	(ⅤD)
teminology	[təmi'nɔlədʒi]	n. 术语，词汇	(2)
temper	['tempə]	n. 回火	(3)
terminal	['təːminəl]	n. 终点站，终端，接线端	(20)
terminate	['təːmineit]	vt. & v. 结束；使终结；解雇；到达终点站	(ⅣA)
thermal conductivity		导热性	(1)
through-feed		纵向进给	(6)
thrust	[θrʌst]	v. [机] 推力；侧向压力；插；猛推	(ⅡA)
tightness	['taitnis]	n. 坚固，紧密	(22)
titanium	[tai'teiniəm]	n. 钛	(1)
title	['taitl]	n. 标题，题目；书名；头衔；称号	(26)
token ring network		令牌环网	(ⅤA)
tolerance	['tɔlerəns]	n. 公差，宽容，容许量；给（机器部件等）规定公差	(18)

transfer mechanism	传输机械装置	(20)
transition [træn'ziʃən]	n. 过渡，转变，变迁；[语] 转换；[乐] 变调	(19)
transmission [trænz'miʃən]	n. 传送，传播，传达；播送	(12)
transparency [træns'pærnsi]	n. 透明；透明度；透明性；透明的东西	(19)
tremendous [tri'mendəs]	adj. 极大的，巨大的	(12)
trimming press	冲拔罐修边机	(20)
tungsten carbide	碳化钨，硬化合金	(ⅣD)
turning centers	车削中心	(18)
twist drill	麻花钻头	(8)

U

ultrasonic [ˌʌltrə'sɔnik]	n. 超声波；adj. 超声的；超音波的，超音速的	(24)
unilateral [ˌjuːnə'lætərəl]	adj. 单边的，单向的	(ⅠB)
upper arm	上臂	(22)
upright drill press (upright drilling machine)	立式钻床	(8)

V

valve [vælv]	n. 阀	(15)
valve-stem seals	阀杆密封	(23)
variable ['vɛəriəbl]	adj. 变化的，可变的，易变的；n. 可变因素；变数	(11)
variable-delivery pump	变量泵	(ⅣC)
variety [və'raiəti]	n. 变化，多样性，种种，品种，种类	(ⅣD)
velocity [vi'lɔsiti]	n. 速率，速度；周转率；高速，快速	(ⅡA)
ventilation [ˌventi'leiʃən]	n. 通风设备；通风方法	(16)
verification [ˌverəfi'keiʃən]	n. 证明；证实；核实	(13)
versatile ['vəːsətail]	adj.（指工具、机器等）多用途的；多才多艺的；多功能的	(13)
vibration [vai'breiʃən]	n. 振动，颤动	(ⅠA)
vice versa	反之亦然	(ⅡA)
videoconferencing [ˌvidiəu'kɔnfərənsiŋ]	n. 视频会议	(ⅤD)
viewpoint ['vjuːpɔint]	n. 观点，意见，角度	(12)
Virtual Robot Technology	虚拟机器人技术	(ⅤB)
viscosity [vis'kɔsəti]	n. <术>黏稠；黏性	(11)
visualization [ˌvizjuəlai'zeiʃən]	n. 形象（化），形象化，想像	(19)

W

watchdog timer	监视计时器 [WDT]	(13)

web [web]	n. 钻芯	(8)
whereas [wɛərˈæz]	conj. 考虑到；鉴于	(5)
with the help of	在…的帮助下，借助	(21)
workstation [ˈwəːksteiʃən]	n. 工作站	(20)
wtisitive drill press	台式钻床，高速手压钻床	(8)
zinc [ziŋk]	n. 锌	(1)

参 考 文 献

[1] 张跃. 机械制造专业英语 [M]. 北京：机械工业出版社，2004.
[2] 徐存善. 机电专业英语 [M]. 北京：机械工业出版社，2012.
[3] 刘杰辉. 机械专业英语阅读教程 [M]. 大连：大连理工大学出版社，2005.
[4] 温丹丽，毕秀梅. 电专业英语阅读教程 [M]. 大连：大连理工大学出版社，2007.
[5] 教育部《机电英语》教材编写组. 机电英语 [M]. 北京：高等教育出版社，2001.
[6] 杨植新. 自动化与电子信息专业英语 [M]. 北京：电子工业出版社，2009.
[7] 别传爽. 机电专业英语 [M]. 北京：北京理工大学出版社，2010.
[8] 张琦，杨承先. 现代机电专业英语 [M]. 北京：清华大学出版社，北京交通大学出版社，2005.
[9] 吴卫荣. 气动技术 [M]. 北京：中国轻工业出版社，2005.
[10] 严俊仁. 机电专业英语翻译技巧 [M]. 北京：国防工业出版社，2000.
[11] 刘瑛，徐宏海，罗科学. 数控技术英语 [M]. 北京：化学工业出版社，2003.